SCIENTIFIC AND EDUCATIONAL JOURNAL OF THERAPEUTIC RIDING

2009

ISSN 1174-6238
ISBN 978-0-9845490-0-9

The Federation of Riding for the Disabled International A.I.S.B.L

SECRETARIAT (as of April 1, 2009)
F.R.D.I.
P.O.Box 134 • Damariscotta, ME 04543
Tel: +44 207 1831 256 • Email: office@frdi.net • www.frdi.net

ISSN 1174-6238
ISBN 978-0-9845490-0-9

The Federation of Riding for the Disabled International A.I.S.B.L

Table of Contents

Table of Contents

Editor's Note

This year's highlight for FRDI was the XIII. International Congress of Therapeutic Riding held in August in Münster, Germany. About 800 participants from 39 countries had the opportunity to hear many excellent presentations and enjoy in hand work demonstrations in all of the fields of TR.

My goal for this journal was to collect student abstracts from those countries that offer education in the broad field of Therapeutic Riding. While in Münster I met educators from many countries and enjoyed discussions and a lively exchange of knowledge. My initial idea, however, only became partially true and you will also find several final reports in addition to the abstracts. I hope you will read this journal with interest. If you would like to receive more detailed information about a specific article, you will find the contact information of the author on page VII.

I hope to see many of you in Stockholm in July at the 2010 IAHAIO congress. Therapeutic Riding will be an important part of the congress schedule!

I would like to extend my gratitude to all those who took part in this project! A special thank you goes to Jennifer Dixon-Clegg for her diligent efforts.

Sanna Mattila-Rautiainen

Vice President FRDI

The Federation of Riding for the Disabled A.I.S.B.L. would like to acknowledge with gratitude the generous contribution of an anonymous sponsor from Belgium. This donation made possible the design, printing and distribution of this publication.

Guidelines for Authors

Federation of Riding for the Disabled International (FRDI) publishes its Scientific Journal annually and seeks submissions of original work for inclusion.

Submitted articles may be reports of research, theory, practice, education, or reviews of existing literature. All articles will be independently and anonymously reviewed. Selected articles will make a contribution to the advancement of knowledge in the field of Therapeutic Riding. Submitted text should be unpublished and should not be under consideration for publication by another journal or publisher. Please contact the editor with questions in this area.

Send us the text in **e-mail format**, Microsoft Word Version 6 or later.

Attach a separate file with your name, address, telephone and fax numbers, e-mail address, along with brief biographical notes (up to 100 words) for inclusion with your article if it is accepted.

An abstract of approximately 100 words should be included at the start of the text.

Tables and graphs should be numbered, referred to by number in the text, and should be placed in text where they are to occur. If possible, try to avoid shading in graphs.

Photographs should be provided separately in JPEG- format.

Acknowledgements should be made at the end of the article, before the list of references.

References:

In the article: Martin, et al. (2002)

In references: Martin, B.L., Sloane, K., & Pearce, N. (2002). Title of article here. Journal Name Here, 2 (23), 234-39.

All correspondence should be sent to the editor at editorial@frdi.net

About the Authors

Article One - Irma Merabishvili

Irma Merabishvili, P. Shotadze Tbilisi Medical Academy, 0170 Tbilisi, Georgia, I. Chavchavadze Ave., 29, P., Phone: (995 32) 25 18 35

Article Two - Daniël Bens

Daniël Bens, president of GISSIG-Hippo, Belgium

Daniël Bens has worked for thirty-five years with children with cerebral palsy and children with psychomotoric retardation. As a neurological and p.m. physiotherapist, he started hippotherapy sessions in 1990. In the beginning, he started with 3 patients a week and today works full time with 60 patients.

Mr. Bens is president of GISSIG-Hippo, a Belgian Interest Group for hippo physiotherapists. He is also a teacher for the Artevelde EAT course and the Para Equestrian courses for trainers. He is an international classifier for FEI Para Equestrian and responsible for the national para equestrian competition.

Article Three - Elina Aula

Elina Aula, Sonasundintie 150, Inkoo, Finland,
Phone: 358-40-5412337, Email: elluaula@gmail.com

Elina Aula is a physiotherapist and hippotherapist. She specializes in work with neurological and trauma patients. Ms. Aula runs her own company working with four horses.

Article Four - Gisa Koller

Gisa Koller, Alleestraße 84, 44579 Castrop-Rauxel, Germany,
Phone: 0049-177-7941422, Email: gisa.koller@gmx.de

Gisa Koller, a university student, is currently working on her final paper. Prior to her studies, she did an apprenticeship in health care and she still works as a nurse. She has been working for Henrike Struck in Lünen for three years where she gained experience in the field of therapeutic riding. At the moment, Ms. Koller is doing an advanced training as a trainer in educational vaulting and riding with the DKThR.

About the Authors

Article Five - Stanislava Tílešová

Stanislava Tílešová, Lipowa 7 97-400 Belchatow, Poland,
Email: stanatilesova@wp.pl

Stanislava Tílešová is a psychotherapist, psychologist, teacher and therapeutic riding instructor. She is Slovakian, lives in Poland where she is a Director of a Therapeutic Riding Centre. She has a Phd in Therapeutic Riding for Children with Behavior Difficulties. Ms. Tílešová has worked with Therapeutic Riding in Germany, Austria, England, Slovakia and Poland.

Article Six - Lovorka Čuljak

Lovorka Čuljak, Faculty of Law, Department of Social Work,
Sv. Mateja 79, 10 000 Zagreb, Croatia,
Phone: 00 385 1 66 047 66, Email: lovorka.culjak@gmail.com

Lovorka Čuljak lives in Zagreb. She graduated from the Faculty of Law, Department of Social Work in 2009 and studied therapeutic riding and the life quality of children with cerebral paralysis.

Article Seven - Margreet Ludewig

Margreet Ludewig and Jean Fourie, University of Johannesburg,
P.O.Box 1467, Allens Nek 1737, Roodepoort, South Africa,
Email: margreet@iburst.co.za

Margreet Ludewig taught for many years in mainstream and special needs education. She has an Honours degree in Inclusive Education and a Masters degree in therapeutic riding as a supplementary support for children with special needs. She works as a remedial riding instructor assisting children with CP, Autism, Down's syndrome and other learning impairments. Jean Fourie is a lecturer at the University of Johannesburg, Faculty of Education. She trains educational psychologists, school counselors and life orientation teachers. Her research is in inclusive education and learning support. She has worked extensively as an educational psychologist in special needs education and consults on implementing inclusive education.

About the Authors

Article Eight - Michał Kaczor

Michał Kaczor, M.Sc. 17b/82 Dzielna Street, 01-029 Warsaw, Poland, Phone: 48 880 256 050, Email: mk1977@tlen.pl

Michał Kaczor is a social psychologist and occupational therapist working with people with disabilities. He is a member of the Polish Hippotherapeutic Association and offers hippotherapy at the Handicapped Children Aid Foundation HIPOTERAPIA. Mr. Kaczor is also Vice-president of the MOŻEMY Foundation.

Article Nine - Petra Nowak

Petra Nowak, Kangasala, Finland, email: petra.nowak@nic.fi

Petra Nowak is a Riding Therapist with a degree in Physiotherapy from Pirkanmaa Polytechnic and trained in the Bobath concept. She has worked in the United Kingdom as well as in Finland, focusing on neurological physiotherapy rehabilitation for adults with severe neurological disorders. She has also worked in intensive neurological rehabilitation for patients with acute CVA, SCI and severe head traumas. Since completing her training as a Riding Therapist in 2007, Ms. Nowak has focused on riding therapy for neurological patients (both adults and children) with a wide variety of diagnoses, such as MS, CP, SCI, Learning Disabilities, rheumatoid arthritis, chronic pain, neuromuscular disease, CVA and trauma induced head injuries.

Article Ten - Susanne Hecht

Susanne Hecht, susi.hecht@freenet.de

Susanne has a Bachelor of Arts in "Rehabilitation Pedagogy" from the University of Dortmund and a Masters degree in "Motology" from the University of Marburg, Germany. Ms. Hecht has done internships in Therapeutic Riding at the Centre for Therapeutic Riding in Lünen, Practice for Therapeutic Riding, Dortmund, Moto-Pedagogical Riding Therapy, Marburg, Community of Interests for Therapeutic Riding, Wehrda, all Germany.

About the Authors

Article Eleven - Lorenzo Lucarelli, Montse Cuesta, María Barajas

Lorenzo Lucarelli, AEDEQ, Spain, Email: lucarellil@hotmail.com

Lorenzo Lucarelli is a therapeutic riding teacher, philosopher, and president of AEDEQ, the Spanish Association for Equine therapies.
Montse Cuesta is a psychologist with AMICS, an Association of Sabadell, near Barcelona, which is dedicated to improving the quality of life for the mentally challenged.
María Barajas is a nursery assistant and student.
AEDEQ and AMICS have actively cooperated to set up a service for the mentally challenged based on the relationship between horses and humans.

Article Twelve - Jane Liddiard

Jane Liddiard, Therapy Focus, Perth, Western Australia,
Email: janevl@live.com.au

Jane Liddiard is an Occupational Therapist working for Therapy Focus, a not for-profit organisation in Perth, Western Australia. Her work is with children who have disabilities. In addition to doing Hippotherapy, Ms. Liddiard has been involved with Riding for the Disabled for over 20 years and is a level 2 coach.

Article Thirteen - Carine Debandt & Stefi Lejeune

Carine Debandt & Stefi Lejeune, Arteveldehogeschool Gent, COMPahs, Campus Kantienberg, Voetweg 6, B – 9000 Gent, Belgium, Email Contact: Godi De Vos: godi.devos@arteveldehs.be

Carine Debandt is a Physiotherapist and an Occupational Therapist.
Stefi Lejeune is an Occupational Therapist.
Both, Ms. Debandt and Ms. Lejoune, are 2009 graduates of the EAT Arteveldehogeschool in Gent.

ARTICLE ONE

Riding Therapy in the Disturbance of Balance and First Degree Idiopathic Scoliosis

AUTHOR: IRMA MERABISHVILI, P. SHOTADZE
TBILISI MEDICAL ACADEMY, GEORGIA

Article One

Riding Therapy in the Disturbance of Balance and First Degree Idiopathic Scoliosis

The goal of this work was to determine the disturbance of bearing and first degree idiopathic scoliosis in relation to age.

The children were divided into two age groups (three groups of children aged 8 to 11 years and three groups of children aged 11 to 15 years). There were 8-10 children in each group. Riding Therapy was performed as monotherapy until the completion of puberty. The whole period of the treatment was divided into 4 stages.

The goal of the first stage was the psychological and physical preparation of the children for riding. The goal of the second stage was the reconstruction of an antigravitational system from standing to the state of horse-riding.

The goal of the third stage was the stabilisation of the nucleus pulposus of intervertebral disc and its correction in the case of deviation from the centre. Against a background of self-stretching through riding to achieve the strengthening of plastic tonus of muscles maintaining the spine, exercises were performed in an isometric regimen, creating a stronger muscular corset around the spine.

Investigations carried out at this stage established that treatment starting for children at the age of 8-9 years was more effective than the treatment started in the second (older) age group. Thus, the Riding Therapy should begin ahead of the pubertal period.

The goal of the fourth stage was to retain a medical effect, to improve general physical development, to increase a psycho-neurological tonus, to prevent the state of attention deficit/hyperactivity in children and to consolidate a sound mode of life.

ARTICLE TWO

Does Hippotherapy Increase Axial Control of the Trunk as a Treatment for Individuals with CP? Are the Results Linked Closely to the Frequency of the Sessions?

AUTHOR: DANIËL BENS, PRESIDENT OF GISSIG-HIPPO, BELGIUM

ARTICLE TWO

DOES HIPPOTHERAPY INCREASE AXIAL CONTROL OF THE TRUNK AS A TREATMENT FOR INDIVIDUALS WITH CP? ARE THE RESULTS LINKED CLOSELY TO THE FREQUENCY OF THE SESSIONS?

For the hippophysiotherapy sessions, the target group was sorted by the topographic cerebral palsy classification: (congenital) diparesis, with mild to moderate spasticity and sensorimotor impairment, characterised by a lack of normal movement experiences.

This study was carried out with the assistance of Jolien, a 7-year-old girl, suffering from a motor disorder called spastic diparesis with a mild mental handicap. For four consecutive years she had some Botox infiltrations near the hamstrings, the last one in January of this year (2009). She moves functionally with an orthopaedic tribike, a manual wheelchair and a Kaye-walker. She wears bilateral leafsprings in orthopedic shoes and a night orthesis. Apart from the typical problems in the lower limbs she suffers from a weak hypotonic trunk musculature (postural and selective) as well as from clear instability near the pelvis region and poor selective muscular strength values for all hip muscles at the expense of the right side.

The working arrangement with the attending physiotherapist is as follows:

Increasing postural and dynamic trunk control with attention to the pelvic region and working in higher positions. The hypotony-induced poor pelvic control in turn increases tension and decreases mobility in the lower limbs.

However, Jolien is rather lazy, both as regards her starting position and the performance of activities. A new challenge therefore had to be sought. As far as techniques are concerned, Bobath gives the following advice: "use what works best". As from now the therapy horse will be involved in the therapy.

The aim of the study was to confirm the efficiency of postural stretching of the trunk to improve control of the pelvis when riding a horse in free walk without a saddle. The three-dimensional movements of the horse back give sensory information to the trunk muscles, which are required to react to the changes in posture during riding and to postural correction. It goes without saying that a combination of both is required to maintain trunk balance.

In view of Jolien's ideal age, (between seven and twelve years), the development of her coordination ability can reach maximum potential.

Tone measurements were made with biofeedback equipment and trunk posture simultaneously registered on tape. In today's environment it is easy to digitalize and arrange the tapes with supplementary information. A comparison is made with the activity of the lumbar erector muscle during everyday activities. The different recordings take place in identical conditions, same location, positioning of the framework and camera, electrodes on measured places. Contrary to, for example, assessment with the GMFM test, measurements are made immediately and on the spot. There is no trial period and no control group, as this has already been illustrated sufficiently.

It is demonstrated that muscle activation during a hippo session is considerably larger than in everyday conditions and it is therefore useful to continue this therapy with a neurophysiologic background.

The starting point is to demonstrate the results of the practice session, at the very moment riding the horse, and in different movement functions and situations.

A tailored activity is required to increase muscular strength. Within this scope we resort to carrying weights, offering resistance and sensory stimulation in several positions to develop a sufficient basis for training in functional and everyday tasks. The evaluation of these targets is monitored by means of measurements, in this case biofeedback and video registration. Initially, the hippotherapy session was given once a week. I will discuss this in more detail later on.

The therapy includes a combination of individual rides and "pillion" rides, (or "back riding"), i.e. rides where the co-therapist sits behind Jolien on the horse. This makes it possible to direct instructions and to maintain focus. Each session ends with an individual ride, i.e. a ride where Jolien sits on the horse alone. Jolien sits on the horse without a saddle, being held if necessary, both during individual and "pillion" rides. The saddle is omitted to allow the movement of the horse to be better felt and to increase the psychotherapeutic contact. Rides are both with and without stirrups. The assistant guides the horse onto the outside track in walk on both reins, regularly making large and small circles and serpentines. Regular provision of breaks ensures especially that head and trunk stabilisation are encouraged so that a steady effect is achieved on the trunk musculature.

Special interest was in the movement transferred to the rider's lumbar and pelvic area when the horse was in free walk.

The horse continues to move at a walk because the horses walk has the same movement pattern as two normally functioning lower (human) limbs. This results in a pelvic tilt and a pelvic rotation linked to lateral movement. This also applies to the lumbar column, which experiences a lateral and rotated movement (Gottwald). Using a wedge cushion allows an intended increase in proprioception, i.e. to accentuate the correct position of the pelvis. This makes it possible to increase the tone of the lumbar muscles. The patient has the opportunity for a postural and movement experience which she would never have experienced in her neurological evolution.

Only a limited number of different exercises are given because we prefer frequent repetition to reach automation, the last stage in motor development. This opens the way for the next step in the physiotherapy treatment, coordination training.

As far as the frequency of the therapy sessions are concerned, I do not have the means or time to perform scientific research, but I can illustrate this aspect of physical hippotherapy via this contribution, which I think is worth sharing.

In the current calendar a year has 12 months, one month has 4 weeks and one week has 7 days, 5 of which are normally working days. Hippotherapy is in most cases given once a week. Suppose the Romans or Napoleon had decided that a week would comprise 11 days, this would have meant that physical hippotherapy sessions would only take place once every 11 days. And we can take this reasoning as far as we like. I furthermore believe that most sessions are currently given once a week purely because of practical considerations. This also applies to the duration of the session, 2 per hour, which is quite convenient. Actually an arrangement without any proper basis. Why do I pay special attention to the frequency of the number of sessions?

In my capacity as physical hippotherapist at an institute for children suffering from neurophysiologic disorders I am also responsible for keeping count of the financial contributions of the patients for the hippotherapy. What does this demonstrate? That even a frequency of once a week is very often not practically possible. Project week, birthday party, holidays, Botox infiltrations, St. Nicholas, Christmas, sports days etc. As we did not make much headway in one year, we experimentally increased the frequency to four sessions a week, including 3 sessions of 20' and one "pillion" ride session of 30', for about 4 months. After 4 months we obtained the same result as we did with Frederic a number of years ago under similar conditions and with the same technique, Hippo fort postural stretching and dynamic trunk control.

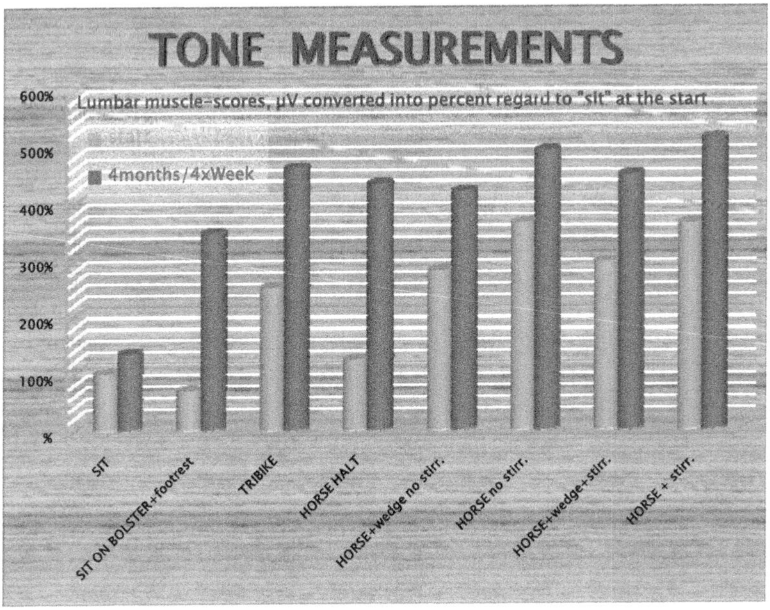

The result in everyday life:

- Better able to stretch when sitting and standing
- able to sit on her own without footrest with lateral support of the upper limbs
- able to sit on her own on a bench with footrest and to clap her hands in diameter above her head
- able to sit on her own on a bench with footrest when performing core functional activities
- better proprioception of stretching
- better postural control with position changes
- improved functionality with orthopaedic tricycle

I think we may presume that an increase of the treatment frequency is of utmost importance to achieve specific results. Results such as higher axial trunk control.

ARTICLE THREE

Possibilities in Hippotherapy with SCI (Tetraplegia)

AUTHOR: ELINA AULA, PHYSIOTHERAPIST AND HIPPOTHERAPIST, FINLAND

ARTICLE THREE

POSSIBILITIES IN HIPPOTHERAPY WITH SCI (TETRAPLEGIA)

BACKGROUND:

Hippotherapy (HT) is a rehabilitation method for neurological clients. There isn't much experience of hippotherapy with SCI clients in Finland. Finnish Riding therapy includes hippotherapy, groundwork, carriage driving, educational riding and vaulting.

Methods: The purpose of the case report was to clarify the possibilities of HT for SCI (tetraplegia) client, and find out what HT can give to the rehabilitation process.

Client: 24-year old female, ASIA B, motor level left C6, right C7. Before a car accident 2003 she was a competition rider.

The data includes 4 weeks of hippotherapy; 3 therapy sessions per week, ground work with horse and carriage driving. Documentation with video camera, camera and the follow up of the client's functions in daily living.

HIPPOTHERAPY

Aims:

- Reduce spasticity
- Improve sitting balance
- Improve sitting posture with the help of aids
- Improve body awareness

Results:

- Spasticity reduced according to VAS-scale, client's subjective sensation before and after hippotherapy. The effectiveness of hippotherapy session was for up to 24 hours.

- Sitting balance with eyes closed improved, from 8 sec before hippotherapy to 53 sec after 12 therapy sessions.
- Sitting posture was supported with elastic belt
- Body awareness improved, the client had better body image and self concept after the hippotherapy period

Groundwork

Aims:

- Interaction with therapy horses
- Confidence between client and horse
- Physical aims: improvement of sitting balance, strengthening of the muscles, working with both hands, trunk rotations, weight shifting in the wheelchair, hand-eye coordination
- To be able to take care of a horse by herself
- Lead a horse on her own

Results:

- The client can take care of a horse and lead horse on her own. Ground work was done with two therapy horses. It was important that the client could trust both horses.

Carriage Driving

Aims:

- The client can drive a horse and carriage and handle the horse independently, like she could before her accident.

Results:

- Carriage driving was possible with a vehicle designed for wheelchair users. Driving gave the same exercise to the client as riding did before the SCI. She could train a horse by herself and work as a couple with the horse.

Conclusion:

This study shows that after tetraplegia, the client can sit on a horse alone, do groundwork and drive a carriage. A trained therapy horse, hippotherapist and special aids are required to perform high quality and safe work with tetraplegic clients.

THANK YOU TO:

Anna, all the helpers and wonderful Icelandic horses

ÞRóTTUR FRÁ SLÉTTU and BRIMIR FRÁ STÓRA-SANDFELLI2

References:

1. **Sandström, Marita (2000)**
   Ratsastusterapian vaikutukset neurofysiologian ja tutkimustulosten näkökulmasta. Suomen Kuntoutusliitto

2. **Strauss, Ingrid (1995)**
   Hippoterapi. Neurophysiological Therapy on the Horse. Ontario Therapeutic Association

3. **Zäch, Koch (Hrsg.) (2006)**
   Paraplegia Ganzheitliche Rehabilitation

4. **Gäng, Marianne (1990)**
   Heilpädagogisches Reiten und Voltigieren. Ernst Reinhardt Verlag

5. **Scott Naomi (2005)**
   Special Needs, Special Horses. A Guide to the Benefits of Therapeutic Riding. University of North Texas Press

ARTICLE FOUR

Single Case Study on the Change of Motion in Adults with Cerebral Palsy by Psycho-Educational Vaulting and Riding

AUTHOR: GISA KOLLER, GERMANY

ARTICLE FOUR

SINGLE CASE STUDY ON THE CHANGE OF MOTION IN ADULTS WITH CEREBRAL PALSY BY PSYCHO-EDUCATIONAL VAULTING AND RIDING

INTRODUCTION

Many people who are affected by severe physical disabilities due to cerebral palsy and spasticity are using the opportunity of psycho-educational vaulting and riding therapy. After riding, those affected, as well as their caregivers, reported positive effects both on their mood and on improvements in mobility.

Even during the promotion, improvement in mobility and tonicity can be observed. Therefore, I was faced with the question of how the change in mobility through riding could be measured, and direct effect represented. In addition to the development of a scientific method for measuring mobility, a measurement is needed for the individual which is suitable for daily use but which does not detract from therapy time.

The following article, which is an overview of my thesis-led single case study, describes the impact of remedial support with horses on the mobility and spasticity of adults with cerebral palsy.

RESEARCH METHODS AND DESIGN

The mobility was measured by using two different methods. As a scientifically accepted method, the joint was measured with a goniometer, also called normal-zero-method. Moreover, an individual diagnostic menu for the group of people was developed. Important features of the diagnostic menu are the fast feasibility of the measurements, practicality and the possibility for individual extension adaption to other participants respectively.

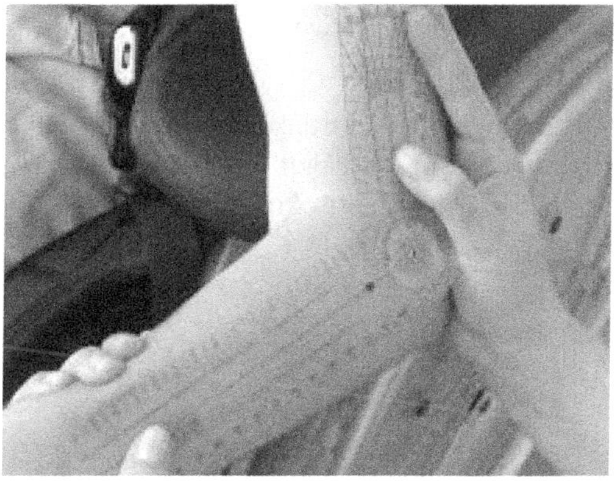

Figure 1: Joint measurements with the goniometer, the left elbow

In addition, the vital signs such as blood pressure, heart rate, body temperature and the surface temperature of the hands were measured. The respective measurements were carried out before and after each treatment unit.

The neutral-zero method offers the ability to document the joint mobility. The starting point for any measurement is a uniformly defined 0-position. This refers to the normal anatomy of each joint, which a healthy person reaches in the upright position with arms hanging down and parallel to the foot (Debrunner, 1978). The mobility of a joint is documented with 3 numbers that specify the two end positions of the joint, and the zero position. Is the reason for the zero position not being achieved, due to the spasticity or contracture? (The mean value represents the maximum extension.)

For this study, extension and flexion of the elbow and the knee were measured using this method. All movements were performed passively. For the mobility measurement with the diagnostic menu, extension and flexion of the cervical spine, abduction of the hip and the overall mobility of the torso with the upper extremities were all measured.

The required materials consist of two clothes pegs, duct tape and a tape measure.

With the tape measure, knee distance (abduction of the hip) and chin-sternum distance (cervical flexion / extension) were measured as well as the output value of the respective measurements to a fixed point on the therapy girth.

To measure the improvement in trunk mobility, at the beginning and end of each therapy session the riders were asked to reach as far up the mane as possible. Their points of reach were marked with a clothes peg.

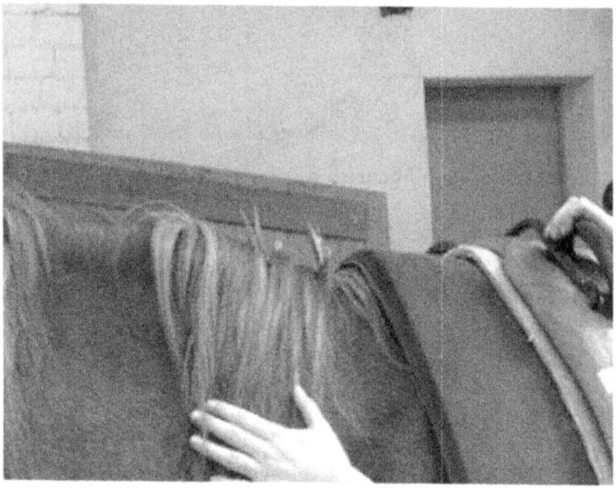

Figure 2: clothes pegs

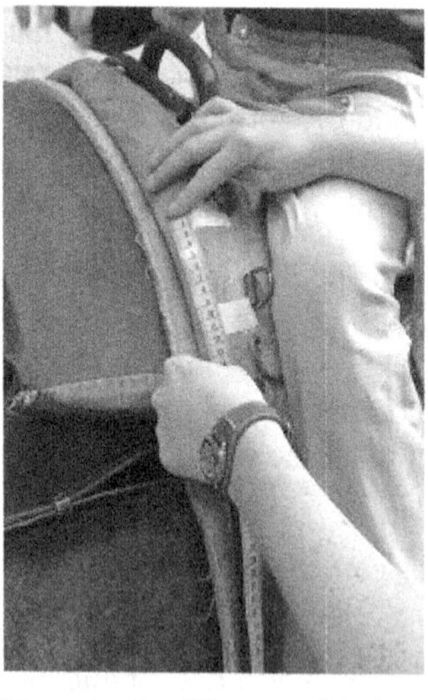

Figure 3: Measuring the difference of the knee position

After mounting and before dismounting the duct tape was used to mark the position of the knee, the seat and the heels. The particular differences, as well as the initial value were recorded to a fixed point on the belt.

Vital signs were collected to see whether and how riding affects the cardiovascular system. Of particular interest was the surface temperature of the hands, because it can indicate an improvement in blood flow to the extremities.

Riders and Intervention

The measurements were carried out once a week over a period of six weeks. As a means of comparable control additional measurements were also carried out on a second person over a two day period.

During therapy, the horse is led, and a second person sits on behind the rider as assurance. Each therapy unit was carried out on the same horse. Treatment time on the horse ranged from 9 to 19 minutes.

Among the riders are two men with severe multiple disabilities due to early childhood brain injury. Rider 1 is quadriplegic, rider 2 (control person) is diplegic. Both are wheelchair dependent, and because of their strongly pronounced spasticity they require constant assistance.

Results

The following graph shows the results of the pre-and post-tests of rider 1. The results do not differ significantly from the results of the control person, rider 2.

NORMAL-ZERO METHOD

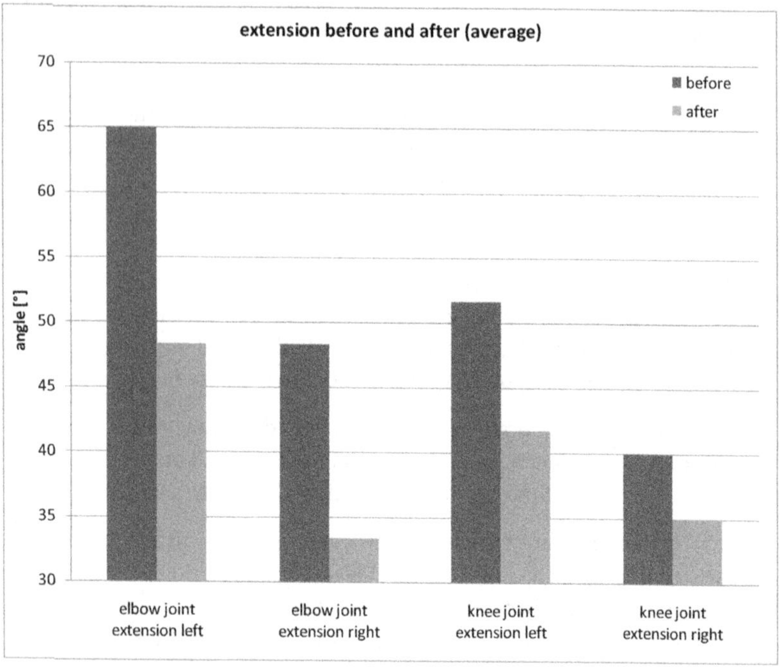

Figure 4: Averages of flexion of the pre - and post-tests on the elbow and knee

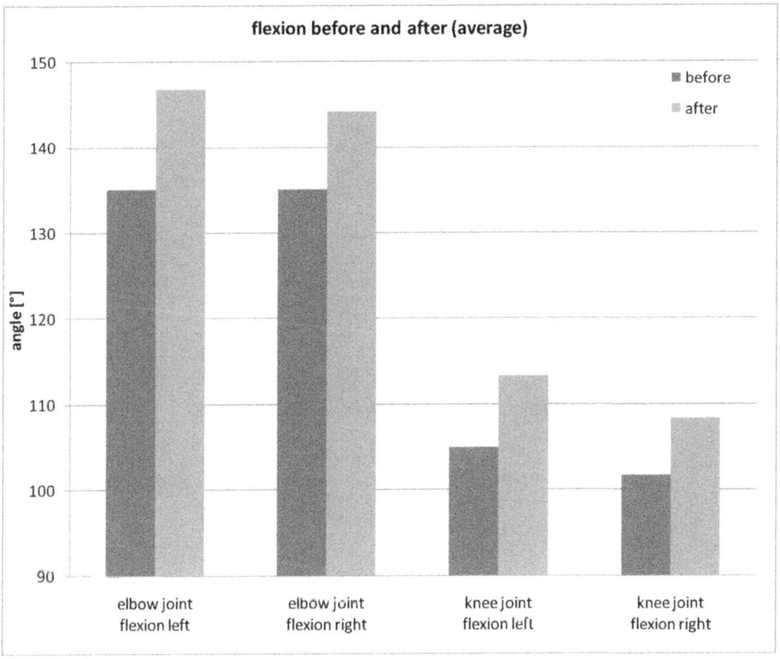

Figure 5: Extension of the average pre-and post-test on the elbow and knee

The figures show that neither during pre test nor during post-test did joints reach zero position.

Due to the significantly poorer results of the measurements on the left side of the body it can be seen that this side is more affected by spasticity than the right. Assuming a standard value of the elbow by 10-0-145, the graphs show that the values of the extension are worse than the inflection. The same applies to the measurements of the knee

It is striking that the changes in the average for all measurements are at 20 degrees, which the following figures show using the example of the elbow joints.

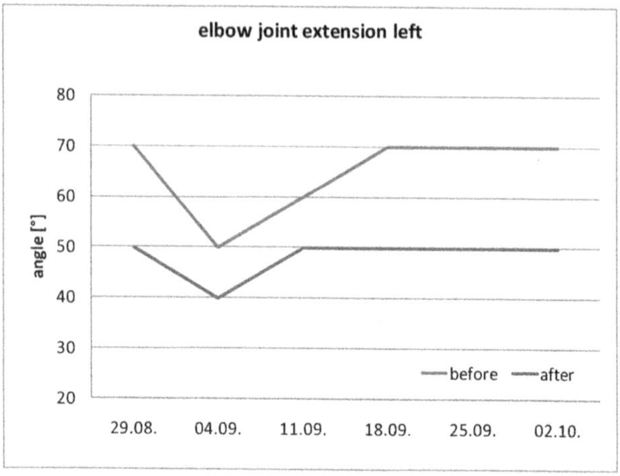

Figure 6: History of the measurement of the left elbow

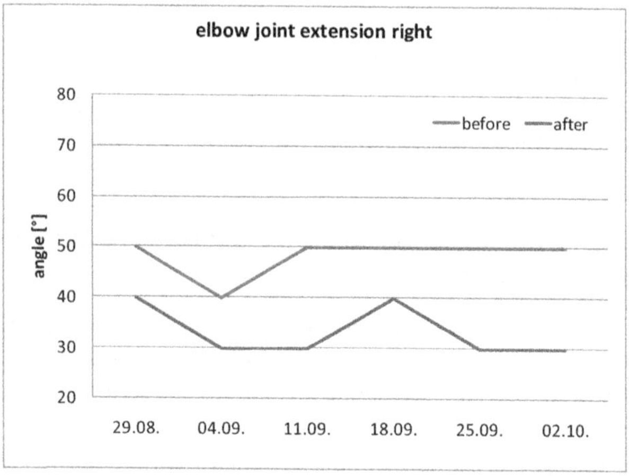

Figure 7: History of the measurement of the right elbow

The two graphs show the measurements of the elbow joints over a period of 6 weeks. It clearly shows that the traces of pre-and post-tests are relatively parallel to each other. The output value of the left elbow on 04 September corresponds to the recorded value of the measurements at 11 September. The output value of the right elbow on 04 September corresponds to the recorded value of the measurements on 18 September.

This demonstrates that, apart from the baseline, an improvement of about 20 degrees is achieved. A change of more than 30 degrees was not reached in any measurement; however an improvement of at least 10 degrees always took place.

DIAGNOSTIC MENU

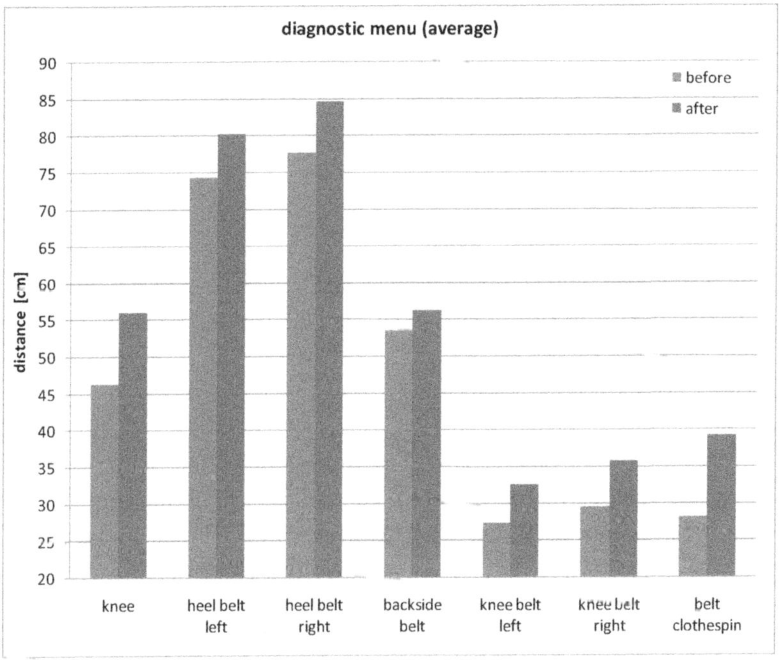

Figure 8: Averages of the entire pre-and post measurements of the Diagnostic Menu

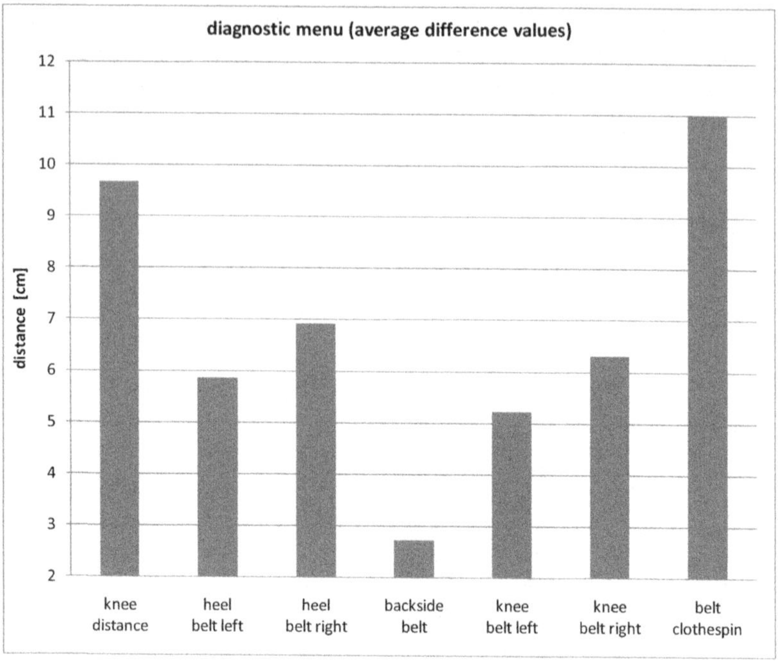

Figure 9: Average difference values (changes) of individual measurements

As with the measurements of the standard zero-method, consistently positive results can be detected from the graphs. The most significant results are the measurements of the distance between the knee and trunk mobility (clothes peg), an average of 9.7 cm, or 11cm was measured (Fig. 9). This finding reveals that the hip and trunk mobility is influenced most positively by riding.

Again it is striking that the differences between the pre and post value do not significantly differ in the diagnostic menu throughout the procedure of measurement. This is supposed to be illustrated by the measurements of the knee over the entire measurement period (Fig. 10). The differences of the values in the pre-and post-tests range from 8 to 12 cm independent of the baseline.

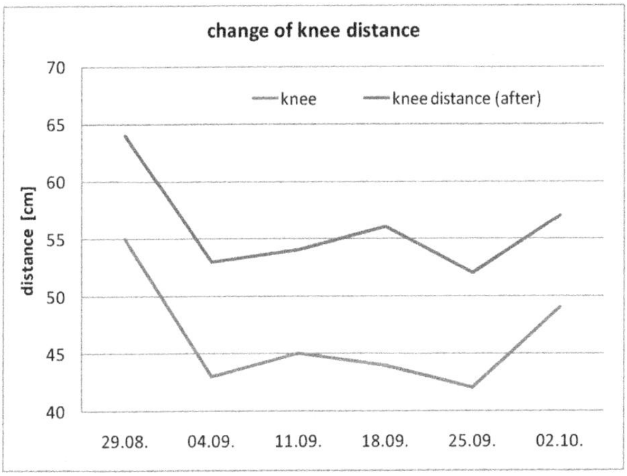

Figure 10: Measurements of the knee distance during the interval

In addition, the time on the horse was documented to determine whether prolonged riding time would improve the results.

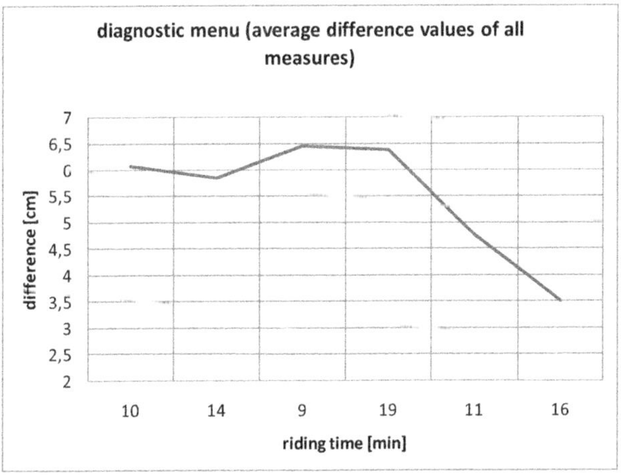

Figure 11: Average difference values of all measurements of the Diagnostic menu in relation to time

The illustration shows that the results of the measurements cannot be related to the period of time. The largest average positive change was measured at the shortest riding time of 9 minutes. The change was 6.5 cm. It can therefore be assumed that the positive effect of riding and the relaxation of spasticity are achieved in the first minutes on the horse.

Vital Signs

Monitoring vital signs showed that the post value of blood pressure, pulse and body temperature did not differ materially from the pre values. The cardiovascular system is therefore not strongly affected by riding. The change in surface temperature of the hands did prove significant. After riding the temperature was much higher than before, indicating an increased blood flow to the extremities, and therefore the muscles.

ARTICLE FIVE

The Model of Variable Factors of the Alternative Emotional Experience in Therapeutic Riding

AUTHOR: STANISLAVA TÍLEŠOVÁ, PSYCHOTHERAPIST, PSYCHOLOGIST AND THERAPEUTIC RIDING INSTRUCTOR, POLAND

ARTICLE FIVE

THE MODEL OF VARIABLE FACTORS OF THE ALTERNATIVE EMOTIONAL EXPERIENCE IN THERAPEUTIC RIDING

The alternative emotional experience (AEE) offers a unique chance as a healing element in therapy with children and adults alike. It often brings a sudden change in the development of the individuals in therapy. Based on the factors of the alternative emotional experience which manifest themselves in the interaction between humans and horses - observed and classified by Kupper-Heilmann - we can set a theoretical model of the work with the clients in therapeutic riding, synthesizing several psychological theories. This model is essentially eclectic, bringing together the outcomes and practical experiences from years of hippotherapy, teaching and my work as a school psychologist. These experiences confirm the validity of the theory of deprivation and its impact on the psychological development.

Developmental psychology shows the importance of the dominant needs in various stages of development, and this is also fundamental with the individual factors of the alternative emotional experience.

Kupper-Heilmann writes about three basic factors of the AEE:

- The experience of being carried by the horse (which includes accepting the client and continuing relationship);
- The experience of a possibility to affect the horse (the possibility to get over one's feeling of helplessness)
- And the experience that the performance of control is necessary and possible (the experience of positive control and the acceptability of limits)

The model of mutual interaction of these factors was created after more detailed analysis, with the help of Maslow's hierarchy of needs (the satisfying of which forms the basis of an individual's development), combined with Erikson's psychosocial theory of development and further theories of moral development (including cognitive development).

We need to realise that physical age is of no consequence here – it is mental age that is important. The individual can be a grown-up based on the successful solution of the fifth stage issues (according to Erikson's eight psychosocial development crisis stages), for example in the work area, but the same person can have problems in personal life which are typical for the first stage.

The alternative emotional experiences stimulate the clients' differentiation between themselves and other people. The AEE experiences come through the closeness of the bodies, without any bias, in the most natural way and are often presented non-verbally.

The areas and factors of AEE are diffuse. We cannot say that we work only with one area of a client's personality.

The experience of competence anticipates responsibility and conscious action. This way AEE becomes the basis of self-realisation. Very often, the client processes these experiences unconsciously.

Interaction of Various Alternative Emotional Experiences

Now let's have a look at how we can deal with some client situations and what consequences this model has for psychotherapeutic work.

Development psychology speaks about particular dominant needs in the individual stages of one's development. The same is true for Erikson's psychosocial theory and it also plays an important role with the individual factors of alternative emotional experiences, according to Kupper-Heilmann.

- The experience of being carried by the horse (which includes acceptance of the client, trust, support and the continuing therapeutic relationship). This experience typically satisfies the need for safety, love and belonging. The same shows in the first stage of psychosocial and moral developments: trust – mistrust (moral anomaly).
- The experience of the possibility to affect the horse (experience of competence, the possibility to get over one's feeling of helplessness, responsibility and creativity) is mainly connected

with the satisfaction of the need for esteem of the first type (power, problem resolution, self-confidence, independence). It can also be connected with the psychosocial and moral development stages: autonomy versus shame, initiative versus guilt (pre-conventional moral reasoning) and industry versus inferiority (conventional moral reasoning).

- The experience that the performance of control is possible and necessary (the experience of positive control and the acceptability of limits) satisfies the need for esteem of secondary type (respect, status, popularity, dignity, respect) and the need for self-actualisation. That is important in the following stages: initiative versus guilt, industry versus inferiority and identity versus role confusion (post-conventional moral reasoning).

Since all the factors influence each other, we need to emphasise the predominant effect of the particular factor in the specific development stages of the individual.

A positive conclusion of the first psychosocial development stage **trust versus mistrust** with the positive outcome **hope** (first year of the life) brings the belief that the satisfaction of needs is possible; so is the possibility to renew the positive relationship with the outside world after experiencing frustration. The child learns which needs are real and which are not.

At this age we use therapeutic riding only with the assistance of a physiotherapist (since there are too many impulses which are too intense for such a small child.) The child mainly needs to harmonise his/her biorhythm with the biorhythm of the mother. The contact with the horse has to be very gentle, must be individually tailored for the child, and the child's mother should be also present.

Older clients who are fixated to this stage need different positions on the horse based on the development level of the brain (positions on the belly, then on the back and later vertical positions). The client needs a lot of warmth, and stimulation through tactile and proprioceptive methods – attention is paid to the vegetative functions of the horse and the client (breathing, heart rhythm...), we use caresses, massage, touch, relaxation, often with closed eyes, holding hands, hugs, etc. The horse is always lead by the therapist, while the client experiences the care he/she needs. Or the therapist is close. The best is not to use a saddle, only a blanket on the horses back, nothing at all or a soft mat.

We respect the limits of the individual client's body and we adapt the rhythm of the horse to the client's needs. Mainly, we use slow or regular step and standing positions. That evokes relaxation. Children usually react very well when they can work together with the therapist. So, for example, we can sit on the horse with the child. There is little need for verbal communication and the therapist's voice should be relaxed and suggestive. A silence is often very well received by the child client. The most important thing is the therapist's relaxed attitude and warmth. We can see a very strong need for protection - the client wants a lot of closeness, touching - and what is also very important at this stage is feedback, e.g. information about positive manifestations.

We do not interpret in this stage. The clients GET what they NEED and that is sufficient.

The child on the picture loves the massages, this way he can feel himself, others, the boy takes, trusts the horse and the therapist, he feels safe, that he exists – he gets physical warmth and acceptance from the horse and the therapist. He gets tactile and proprioceptive stimulation, the same as the child needs in the first year of its life.

Therapist (leading the body of boy) as a cover for this boy by brushing off the horse.

Progress shows when we observe how the client behaves with the horse. The most important are the wishes of the client and the feedback. We need to pay attention to the feelings and also to the horse's behaviour. The therapist has to work as a mirror. The clients choose those horses from the herd that symbolise the emotionality of the mother and the protection of the father. The anxiety is typical for this stage. **We slowly give the client more responsibility**. The therapist is here to ensure the safety; the client **wants to play and to feel that the therapist and the horse care about him/her**. When progressing to the next stage we can change the choice of the horse. **As the client manages to take the whole responsibility for the horse we can talk about the progress to the next stage of the development**.

In the age of 1 - 3 years a child develops basic conscience and self-regulation. This is the stage of pre-conventional moral reasoning development and the basis for responsibility, competence and forming an identity. This manifests by the successful solution of the crisis **autonomy versus doubt and shame**, the positive outcomes of this crisis being **willpower and self-control**, the ability to implement one's free choice and control and to realise one's own self. The development of one's willpower is directly connected with the increasing ability to pay attention, manipulate things, and with independent movement and speech. The child follows the expected outcomes and consequences and does not yet realise the meaning of and the need for control and self-control. *At this stage, it is most important to satisfy the need for love and safety, although the need for esteem begins to manifest as well. It means that the therapist has to work with the second factor of alternative emotional experiences and utilise the possibilities for the independent work of the client and exercising his/her influence on the horse.* In the case of the client's handicap there comes the first realisation of one's own difference.

Now, the child is on the horse together with the mother or with the therapist. The most important experiences at this stage are the feelings of success and sensual stimulation, practicing coordination, trying out various possibilities and experimenting with different things. The limits have to be very clear and fair. We have to enable the client to experience the feeling of responsibility and creativity. We practice and support the assiduity. We support the child in dealing with problems on his/her own, independent decision-making (giving and expecting free choice) and exercising control over the horse. For the older clients it is good to work in the saddle. We try to respect the client's wishes – we find out what they would like to do with the horse. *The encouragement and requesting that they assume responsibility play a really big role here.* **Once the client comes with suggestions about how the therapy should look and about what he/she would like to try next, he got to the next stage**.

The child has problems accepting others near the pony – the willpower in negative solution.

After experiencing constructive decisions she cooperates in the group and stops beating others close to the pony.

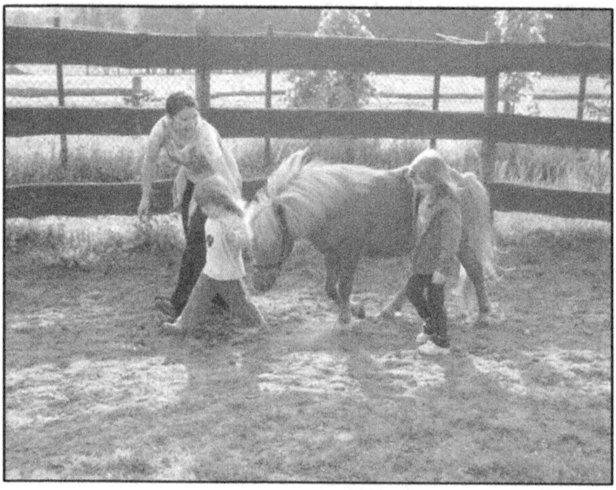

She practices decision making and decides to help

In the stage of the psychosocial crisis **initiative versus guilt**, the positive outcomes are **purpose and direction**. This stage comes in the age of 3 – 6 years and is characterised by gaining self-confidence and ability to make decisions about one's goals without fear and following them through. These qualities develop through play. The child anticipates the situations, problems and roles. The superego is formed. By the end of this stage the child should start feeling guilt when he/she does not follow the rules. A certain level of self-control helps building self-confidence and supports the feeling of independence.

At this stage the child begins to develop pro-social behaviour. This is also connected with a certain level of empathy and ability to manage one's aggressiveness and actual needs. The children with unsatisfied needs for safety and love do not generally feel sure of themselves and often behave anti socially. The acceptance of basic social norms and the ability to control one's behaviour are connected with the continuing emancipation. If the child still feels a strong attachment to the mother, he/she does not feel the need to know the rules. Striving for self-regulation manifests through trying harder to anticipate what may happen and through setting desirable goals. The ability for self control is connected with the development of self-concept (self identity). The child begins to understand the need to adapt his/her behaviour. The child needs to endorse his/her existence with the action that gains him/her credit, while the action itself is not important. What is of consequence is the opinion of some subjectively important authority.

Here, it is very important to work with the second and third factors of the alternative emotional experience. The need for self- fulfillment is satisfied through positively credited actions. The way of satisfying this need turns into a more permanent personality attribute (basic strategies connected with certain values). The child realises that respect for social rules helps his/her inclusion into society. And through that, he/she gets status and a social role. At this stage, the client experiments with his/her possibilities. To support this process we can use didactic aid and try different approaches to stimulate creativity, responsibility and independence.

The child: "What are you doing?"
The therapist: "I am just waiting for you to make a decision about what you want to do." (is crouching 4 meters from the horse, looking once to the boy once to the ground. Makes one's self small and passive)
The horse: "Dear, what is going on here?"

Boy was ready to make the step in the direction of leading the horse. But he was afraid to do it. The therapist's reaction as described (above) provoked him to the experience of own purpose. In 20 seconds the boy led the horse without any problem.

Another boy had problems with hyperactivity and attention. One day he was waiting for his ride on the horse with other children. In just a minute we realised that he was not with the group. Suddenly he started to shout: "Look at that, look at that, I have drawn the head of the horse in the sand." This head was big - at least four meters - and very nice. However he was usually punished for such kind of his behaviour. Instead of punishing him we appreciated his drawing. Suddenly he told me: "I decided to be a good boy. I won't be a bad boy anymore." And it really happened.

The next stage is the psychosocial crisis **industry versus inferiority** with the positive outcome of **competence** (6 -12 years of age). This is characterised by the need to test one's abilities and possibilities. At this age children start attending school and this jump-starts an intensive development of problem-solving strategies. That is the stage of a so-called “good child” in the sense of moral development (conventional moral reasoning). In this process of self-valuation, the moral behaviour becomes increasingly important and forms a part of the child’s identity. The decentration of thinking and the growing importance of social relationships affect development of understanding the moral rules. By the end of this stage the child begins to feel a sense of duty. The children with problems realise their own disability at this time. *The second factor of AEE gives the client a feeling of trust in a competent behaviour. The third factor is important for the feeling of responsibility*. That way we support the resolution of this stage.

The group work in hippotherapy is often crucial. The group serves as a catalyst. The horse clearly reacts to what it likes and what it does not like and in this way the clients can experience the sense of limits and at the same time feel credited for a correct approach. We have to remember to give feedback all the time i.e., to translate the behaviour and feelings of the horse, of the therapist and the other members of the group.

If we work with the parents, teachers or siblings, we have to serve as a translator for all of them. We underline positive interaction. In this phase, the horse becomes a partner to the client (Russig). We give the clients various roles, e.g. an assistant or a sheriff. We have to help them experience that they have various possibilities. We can combine individual work with group work. We let the client care for a horse to the maximum extent possible. The therapist serves only as a translator. The communication between the client and the horse becomes the priority. By explaining the feelings and behaviour of the horse we pave the way for understanding the feelings and behaviour of themselves and other people.

The techniques of a natural communication with the horse are very helpful for the clients to help them understand the necessity of self-control, the need for initiative and exercising one’s will.

One example of giving clients a feeling of competence is for the therapist to go last with the clients leading and deciding.

The next psychosocial stage **identity versus role confusion** with the positive outcome **fidelity** (in the age of 12 – 20 years) is characterised by striving to be true to one's own values. It is a time of big change, searching, experimenting, uncertainty and challenging authority. The need for safety and belonging is often reactivated. Therefore, we need to come back to the first factor of alternative emotional experiences and also satisfy the need for esteem of the second type and the need for self-*fulfillment*; i.e. factors two and three of AEE.

The influence of the horse is very important here, because the therapist is the symbol of authority. We often have to use the techniques catering for the dominant need for safety and belonging. We combine different activities with the horse; we use relaxation, imagination and the models of identification. We have to find the strong characteristics of the client. The children come with the models of identity, e.g. Indiana Jones, and we have to support the strong characteristics of the models. At this stage, the horse becomes the symbol for the client (Russig). That is why the choice of the horse is so important for the client, as it carries the strong characteristics of the identified symbol. *However, full development of the post-conventional moral reasoning stage can probably be observed only with a person in the stage of self-fulfillment.*

Being alone with the horse means feeling closer, being able to trust through self-fulfillment.

Sometimes supporting negativism is important:

One boy with a dominant father was submissive and he couldn't show his own wishes, ideas, feelings. First we did the techniques of horsemanship and then I had one day the need to bring all of the whips that we had on our farm. There were 10 whips and the largest was 4 meters long. I asked the boy to choose one of the whips. He chose the longest one. So I asked: "What can we do with that?" It was the first time that he really looked at the father's eyes and he said: "Let's beat the father!" The father wanted to take the whip from him immediately, but I didn't allow him to. I asked the father to take another one of the whips. So they tried to slam with the whips. The boy was successful, but the father had problems with that. So the next reaction was again the will to take the whip of the son. Again I didn't allow him. Instead I asked the boy to slam louder than the father. The father had to choose a new one of the whips. After this experience the boy started to speak about his wishes, ideas, feelings.

Psychotherapist Tammy Ronen (2000) says that the level of self-control forms the basis for any real therapeutic impact. This can be achieved through play and, according to Bandura (in Hull, C. S. – Lindzey, G,1999) can be developed through the positive example, approval from the others, understanding the good and bad, speech development and realising one's own abilities. That's why the moral development is valuable in this model. It is part of one's identity.

The model shows that we can satisfy the dominant needs by fixation to some stage of development and we can slowly progress with the client to an adequate stage. It is not important whether we work with the past or with the principle "here and now" – the needs of the client should be evident. It is not important how the problems arise but how and why they are sustained. The therapist needs to use his/her intuition about this and to be able "to live" with the particular therapeutic situation. We have to remember about the interpretation here and now. We don't need to think psychoanalytically. Yallom as a very popular psychoanalytic psychotherapist writes after years about his experiences: "The clients remember the therapist's reactions, our way of smiling, laughing, mimic, our human reactions, but not our interpretations!!!" They help us.

We can use the techniques which we use in particular stages for dominant needs for other stages too. The therapist's comments can help to satisfy a client's particular need. For example, leading of horse means:

- for the child who needs the feeling of own competence ("Good job", well done, I like it how you done it")

- but if another child in the same situation needs the warmth of the horse we can say ("This horse really likes you, look at his relaxed attitude! It is happy with you!).

And of course the dominant needs manifest in different ways, they can change in different situations. We should forget the theory and listen to client's health potential. We always have to bear in mind the current situation with the client.

TABLE: THE MODEL OF VARIABLE FACTORS OF THE AEE IN THERAPEUTIC RIDING

Factors of the alternative emotional experience (Kupper-Heilmann)	The Hierarchy of Needs (Maslow)	Psychosocial Stages of Development (Erikson)	Stages of Moral Development (Kohlberg) Implicit Cognitive Development (Piaget)
1. The experience of being carried by the horse	Dominant Needs - group B Safety Love/Belonging	Hope (0-1)	Pre-conventional moral reasoning
2. The experience of the possibility to affect the horse	Dominant Needs - group ? Esteem Type 1 (power, problem resolution, self-confidence, independence) Type 2 (respect of others, status, fame, achievement, dignity, respect by others)	Willpower (1-3) Purpose (3-6) Competence (6-12)	Conventional moral reasoning
3. The experience that the performance of control makes sense, is possible and also necessary	Dominant Needs - group D Self- fulfillment Meta Needs aesthetics, understanding	Fidelity (12-20)	Post-conventional moral reasoning Moral idealism (Muszynski)

Source: Tilesova 2003

BIBLIOGRAPHY:

ČÁP, J., MAREŠ, J. (2000) Psychologie pro učitele. Praha, Portál

HULL, C. S., LINDZEY, G. (1999) Psychológia osobnosti. Bratislava, SPN

JAKABČIC, I., POŽÁR, L. (1996) Všeobecná patopsychológia a patopsychológia mentálne postihnutých. Bratislava, Iris

KUPPER, HEILMANN, S. (1999) Getragenwerden und Einflussnehmen. Groβ-Umstadt, Psychosozial -Verlag

LANGMEIER, J., BALZAR, K., ŠPITZ, J. (2000) Dětská psychoterapie. Praha, Portál

PREKOP, J. (1994) Malý tyran. Co vlastně děti potřebují? Praha, Portál

RONENOVÁ, T. (2000) Psychologická pomoc dětem v nesnázích. Praha, Portál

RUSSIG, D. (2000) Entwicklungsorientierte Arbeit mit Menschen und Pferden. Číž, Hipporehablitačná konferencia.

SERFONTEIN, G. (1999) Potíže dětí s učením a chováním. Praha, Portál

TÍLEŠOVÁ, S. (2001) Hippoterapia u viacnásobne postihnutých z pohľadu psychológa. Diplomová práca. Katedra psychológie a patopsychológie Univerzity Komenského

TÍLEŠOVÁ, S. (2001) Informatio hippoterapeutica Slovaca

TÍLEŠOVÁ, S. (2002) Hippoterapia u detí s externalizujúcimi poruchami správania. Dizertačná práca. Katedra psychológie a patopsychológie Univerzity Komenského

TÍLEŠOVÁ, S. (2003) Teoretický model práce PPJ. 3. hipporehabilitačná konferencia Číž

TÍLEŠOVÁ, S. (2003) Hippoterapia u detí s externalizujúcimi poruchami správania. 3. hipporehabilitačná konferencia Číž

VÁGNEROVÁ, M. (1999) Psychopatologie pro pomáhajíci profese. Praha, Portál

VÁGNEROVÁ, M. (2000) Vývojová psychologie. Praha, Portál

VÁGENROVÁ, M., HADJ, MOUSSOVÁ, Z., ŠTECH, S. (2001) Psychologie handicapu. Praha, Karolinum

ZELINA, M. (1996) Stratégie a metódy rozvoja osobnosti dieťaťa. Bratislava, Iris

ARTICLE SIX

Therapeutic Riding and the Life Quality of Children with Cerebral Paralysis

AUTHOR: LOVORKA ČULJAK, FACULTY OF LAW, DEPARTMENT OF SOCIAL WORK, CROATIA

ARTICLE SIX

THERAPEUTIC RIDING AND THE LIFE QUALITY OF CHILDREN WITH CEREBRAL PARALYSIS

Although the concept of rehabilitation and recreation of people with certain disabilities is still rather unknown, it is becoming more and more popular. Information in media indicates that therapeutic riding has a positive effect on the rider's quality of life.

According to that, the general goal of this research was set to examine the effect of therapeutic riding on some life quality aspects of children with cerebral paralysis, who are attending a programme of therapeutic riding in the association Krila in Zagreb.

Additionally, more specific goals were formed: to examine how parents evaluate the effect of therapeutic riding on physical, psychological and socio-educational aspects of their children and also examine their opinion regarding therapeutic riding and their satisfaction with the programme.

The research was conducted during July and August 2009 on respondents in Zagreb and surroundings. The questions were composed as a survey questionnaire. The respondents were parents of children who attend therapeutic riding in the association Krila in Zagreb (N=25). Most of the children (84%) that were part of this research had been diagnosed with spastic cerebral paralysis, while according to topography, the most represented forms are diparesis and quadriparesis. Additional impairments that appear most frequently are speech impairment, attention disorder, sight impairment, learning difficulties and salivation. 80% of observed children in the association Krila are attending a hippotherapy programme, while the rest of them are involved in a programme of therapeutic riding as a form of sport and recreational rehabilitation.

The results of this research indicate that parents consider that therapeutic riding has a very high positive impact on all observed aspects. Parents' evaluations show that therapeutic riding has a very high impact on sensory integration, strengthening and agitation of torso and pelvic muscles, rigid muscle stretching, spatial orientation, ability to change positions, body posture, sitting, regular head movement, dynamical balance, hand-to-hand coordination and eye-to-hand coordination. It was also indicated that therapeutic riding has an extremely high impact on good mood, interest for external world, self-esteem, life control, confidence, acquisition of new sensory experiences, self-confidence, sense of normality, concentration, the ability to undertake risks, character recognition, spatial perception and perception of shapes. Furthermore, parents have evaluated that as a result of attending therapeutic riding their children have developed love for animals, responsibility towards other people, and sense of teamwork, while as a result of increasing social contacts that children experience during group therapeutic riding they develop sense of friendship. Therapeutic riding also has a large impact on development of speech and communication skills, vocabulary development and differentiation.

None of the parents noticed any adverse effects of this programme. Most of them agreed that therapeutic riding is a good addition to other rehabilitation methods and also is a significant method of rehabilitation in itself. More then half of them consider therapeutic riding as a good source of entertainment. However, 92% of parents consider that their children do not ride enough. Currently there are over a hundred attendants, mostly children, involved in this programme, and 115 of them are on a waiting list. Some of them do not have the possibility of riding therapy offered in their place of residence.

Therefore, it is necessary to constantly educate a number of competent professionals and to continuously promote therapeutic riding, in order to attract a greater number of good people - volunteers. In order to improve the life quality of children with cerebral paralysis, therapeutic riding should be equally included among other rehabilitation programmes. Results and evaluation of the programme should prove its efficiency in rehabilitation of individuals with disabilities.

Article Seven

Therapeutic Riding in the Development of 'Core Skills' in a Paraplegic Preschool Child

Co-Authors:

Margreet Ludewig and Jean Fourie,
University of Johannesburg, South Africa

ARTICLE SEVEN

THERAPEUTIC RIDING IN THE DEVELOPMENT OF 'CORE SKILLS' IN A PARAPLEGIC PRESCHOOL CHILD

ABSTRACT

Various forms of therapy such as physiotherapy, occupational therapy and remedial therapy are available to assist children with learning and developmental delays.

RESEARCH QUESTION:

This study explored therapeutic riding in the development of 'core skills' in a physically disabled preschool child, known pseudonymously as 'Amy'. Amy became a paraplegic while still in her mother's womb. When Amy's mother was seven months pregnant, she was shot through her stomach by hijackers. An emergency caesarean was performed and Amy was born two months prematurely. As the baby began moving, it was noticed that her legs remained immobile. After further tests it was discovered that the spinal nerves close to where the bullet had entered, were damaged. Consequently, Amy was left with no mobility in her legs and she was unable to crawl and has never been able to walk. Owing to the limitations in Amy's movement ability, certain developmental areas required for school readiness were delayed. Her perceptual-motor, socio-emotional and language skills were poorly developed and these weaknesses could be enhanced through the movement modality of the horse. A programme of therapeutic riding was implemented over a year, with the specific goal of stimulating both physical and psychological core skills. The horse, with its distinctive three-dimensional movement, facilitated this unique form of intervention. The research question thus focused on how therapeutic riding enhanced Amy's 'core developmental skills.'

Research Method:

A descriptive case study design was followed. Qualitative data in the form of in-depth, semi-structured interviews was collected from the riding therapist, Amy's mother and her educators. The interviews were transcribed, coded and meaningful units were categorised into themes. Data from Amy's school reports, the occupational therapists report and her drawings were also analysed. Triangulation across the different sources of data ensured that the results were trustworthy.

Research Results:

After receiving riding therapy for a year, Amy showed improvement in many areas of her core skills. Amy's balance and posture showed marked improvement which indicated increased trunk and shoulder-girdle control. She could now 'cross the midline,' catch a ball and remain buoyant whilst swimming. Riding on the horse gave her the opportunity to experience three dimensional space which stimulated the vestibular system and this improved her spatial perception of her own body in relation to the world around her.

Having Amy seated on a horse, as opposed to continually being in her wheelchair, brought about an increase in her body awareness. Her self-drawings showed a change from initially sitting flat on the floor with splayed out legs, to now showing a standing, upright position. Her concentration and listening skills improved in the school classroom as well as during the riding intervention.

Amy showed improvement in her social and emotional development. From initially exhibiting angry temper tantrums, she became more even-tempered, confident and able to express her feelings more appropriately. Her sleeping patterns also improved. She was able to hold longer conversations with her friends at school as she spoke enthusiastically about her riding.

Conclusion

This study concluded that riding is an effective therapeutic intervention for paraplegic children. Even though this study explored riding therapy with one child, the results show that riding therapy would be beneficial for other similar children.

ARTICLE EIGHT

Behavioural Therapies in Riding Therapy Practice

AUTHOR: MICHAŁ KACZOR, M.SC., THE HANDICAPPED CHILDREN AID FOUNDATION HIPOTERAPIA, POLAND

Article Eight

BEHAVIOURAL THERAPIES IN RIDING THERAPY PRACTICE

ABSTRACT

Introduction

Many times in our therapeutic practice, we deal with some problematic behaviour of our patients, which may not be related to the main (health) problems of our patients and are not resulting from patient's disease entity. These burdensome behaviours can disturb the process of our therapy, so dealing with these behaviours is crucial for the whole therapy that we carry on. Therapeutic influence is closely related with having a good contact with our patients. Thus there is a place for behavioural therapies.

Method

Often we cannot diagnose and all the more change environmental factors which influence our patient in his daily life, and which can be a reason of his negative behaviour. However, we can stipulate on the patient desirable reactions in certain situations that may take place during hippotherapy sessions. To cope with negative behaviours of my patients, as well as to achieve intended therapeutic effect, I have included in my hippotherapy sessions some elements taken from therapies based on behaviourism.

In my article I will introduce some of the basics of behavioural psychotherapies and I will present how we can use it in therapeutic riding.

Results

Behavioural techniques which I applied in my hippotherapy sessions proved to be an effective method to correct patient's problematic behaviours. It also positively affected the quality of the relationship between the therapist and the patient.

Conclusion

Behavioural therapies, due to their simplicity and usage of primary learning mechanisms that are common to people as well as animals, can be used with success in therapy of young children or children with considerable cognitive deficits. They are easy to apply and can become a successful element of hippotherapy sessions.

ARTICLE

In this article, I would like to bring closer specific situations when behavioural therapy could exist in hippotherapy. I will introduce some of the basics of behavioural psychotherapies and I will present how we can use it in therapeutic riding.

Many times in our therapeutic practice, we deal with some problematic behaviour of our patients, which may not be related to the main (health) problems of our patients and are not resulting from patient's disease entity. These burdensome behaviours can disturb the process of our therapy, so dealing with these behaviours is crucial to the whole therapy that we carry out. I have been looking for solutions that:

- In a relatively short time will bring effects that can be observed in changing a behaviour of a patient,
- will be easy to apply for every therapist (not only psychologist),
- will be easy to apply at the same time with the main therapy,
- will bring permanent changes in a patient's behaviour by eliminating his negative attitude.

The place for behavioural therapies:

Behavioural therapies deal with those outward behaviours which we can observe. The **advantages** of these therapies are:

- verbal communication with patient is not necessary,
- short-term therapies,
- the educational character of therapy; we can correct disturbances that result from earlier learned wrong behaviours,
- focusing on one, specific problem,
- with the goal of changing the negative behaviour of patient.

(1) Most therapies are appropriate only for children with whom verbal contact is possible. Verbal contact with patient is not necessary. Therapies based on behaviourism seem to be an exception ...they can be used in therapy for patients with verbal contact difficulties.
(2) Traditional psychotherapy is a long-term process. In comparison, the behavioural therapy already gives positive results within several months. (3) Behavioural therapies apply to those disturbances in behaviour which are conditioned by factors related to learning. Negative and positive behaviours are obtained in the same way , as a result of processes conformed to universal laws of conditioning and social-learning. (4) Behavioural therapy focuses on a concrete problem, we do not search for hidden reasons for negative behaviour of patient, because the most important is what is possible to observe, what we see (problematic behaviour). (5) Our purpose is to cause changes in behaviour of patient, relating to, for example, behaviours, emotions or manner of coping with problems.

The base of behaviourism consists of two mechanisms: CLASSICAL CONDITIONING and OPERANT CONDITIONING. **Classical conditioning** has been described as first by Russian physiologist, Ivan Pavlov. Classical conditioning consists in associating of two stimuli: (1) a new **neutral stimulus** (in my example presented in the Table 1 it is view of determined tree) with (2) **a well-known stimulus** that already evokes some organism's reaction (in my example it is reminding about greeting with the horse). Then a new stimulus (conditional) takes over the function from old stimulus (unconditional)! Classical conditioning mechanism acts not only in area of physiology, but also in area of animal's and people's behaviour.

Table 1. Classical conditioning – teaching a child to greet with the horse: associating the command to greet with the horse with stopping the horse by determined tree
unconditional stimulus → unconditional response (Reminding about greeting) (Greeting with the horse) + conditional stimulus → conditional response (View of determined tree) (Greeting with the horse)

The second basic mechanism in behaviourism is operant conditioning. **Operant conditioning** is a learning of a relationship between behaviour and its consequences. Behaviours rewarded frequently bring satisfaction and are remembered; whereas, behaviours that are not rewarded, are smothered and disappear. Therefore, the mechanism allows free reaction, but only one of the behaviours brings a satisfying consequence. In this situation (described in my example) only after word "hup" the horse will ride/move. Patient can choose what to say, but only one, specific answer will be rewarded.

Table 2. Operant conditioning – teaching a child to move a horse:

Only one of the behaviours brings a satisfying consequence

Stimulus („If you want to ride a horse, say "hup" to the horse.")

↙	↓	↘
response A	response B	response C
("Yes")	("Go!")	("Hup!")
↓	↓	↓
no reward	no reward	reward

Speaking of operant conditioning, we should also mention about important principle called **„Premack principle"**. It says that: "Every activity that our patient prefers could be used in therapy as a reinforcing factor for expected, desirable behaviour of our patient." Applying this principle, we should remember to give a reward immediately after the execution of task, but never before!

This was the basis of behaviourism. Now I will introduce the assumptions of these behavioural therapies, which I included in my hippotherapeutic sessions.

The first group are therapies based on counterconditioning. **Counterconditioning** is a strategy of neutralising negative feelings associated with harmless stimuli, like fear of spiders. It uses mechanism of **classical conditioning.** This method involves techniques like **systematic desensitisation** and **flooding.**

The first strategy based on counterconditioning is **systematic desensitisation**. The nervous system cannot be relaxed and stimulated at the same time, because two opposite processes cannot be activated at the same time! The point of this strategy is that a relaxed patient gets closer and closer to anxiety stimulus. Gradual confrontation of relaxed patient with stimulus arousing his fear... reduces intensity of this stimulus. Knowing that two opposite processes cannot be activated at the same time, we try to create the situation when child, who for example is afraid to trot, will be relaxed before riding.

Example 1 - Systematic desensitization

Iga, 3 years old, mild mental retardation.

Iga has never been afraid to trot or learn to ride. Nevertheless, changing her sitting position on the horse was something terrifying for her. I managed to persuade her to sit backwards on the horse by means of... her favourite toy and her father. Together with her father, I started to hide Iga's favourite toy on the horse. Iga was looking over the whole horse for her hidden toy including behind her back. To reach for a toy, which was hidden near the horse's tail, her father encouraged her to sit sideways. With her father beside her, she felt safer. However, this position was uncomfortable for her, so she moved and sat backwards to play with the toy. Feeling safe in her father's presence and satisfied from playing with a favourite toy, caused her to relax and she forgot about fear and sat first sideways and then backwards on the horse. We have been repeating that situation many times and now the terrifying stimulus expired – sitting sideways or backwards is not a problem for Iga anymore.

Flooding is another strategy based on **counterconditioning**. In that therapy the patient is forced at once to face up to his fear, his phobia. However, it happens in a safe therapeutic situation. Multiple repetition of a situation of confronting the patient with **anxiety stimulus** causes extinction of **anxiety power** of this stimulus.

Example 2 - Flooding

Jacek, 35 years old, moderate mental retardation.

Jacek is very short and stocky, but he is very self-confident. He decided to get on a horse for a first time at the western picnic. I was to secure him while getting on a horse and riding. Unfortunately, the horse was quite tall and Jacek is not too much adroit. There was only a small stump available, so getting on the horse turned out to be a big challenge to Jacek. He hung on the left side of the saddle. He panicked and told me to pull him off quickly from the horse. Later, when he saw his colleagues getting on that horse, he decided to try again... and face up to his fear! This time I knew what to expect, so getting on the horse was much easier. When Jacek saw how high he was sitting, he started again to be afraid. I explained to him in detail how I would secure him. Jacek calmed down a little and we started in walking pace. Even though he I was stil tense and he was controlling where I was. At the end of riding, he was more relaxed and even started to pat the horse's neck. At the end, he hugged and kissed the horse. Jacek found himself in **phobic situation**, but earlier I had told him what I intended to do and he had agreed. At the same time, I ensured him a feeling of safety, by the agreed way of securing.

The second group of therapies that I used to include in my practice, were therapies based on operant conditioning. **Operant conditioning** is learning about relationship between behaviour and its consequences. If there is a reward given after some behaviour, that behaviour will be remembered. If there is no reward, that behaviour will disappear. It allows a patient to react freely, but only some specific behaviour brings satisfying consequences. There are two basic strategies of managing by means of consequences, which I used to use in my practise (contingency management): first is the strategy of **positive reinforcement** and the second is the strategy of **extinction**.

Behaviour will be repeated, when it will be immediately rewarded. Frequency of that behaviour will rise. This is the most important principle of **operant conditioning** and the base of strategy of **positive reinforcement.** It is probably the most often applied behavioural strategy in hippotherapy. Every time a patient executes an exercise well and you say: "You did it very good, bravo!" , you are using exactly that strategy. The most important, in this strategy, is that the reward (even very small and insignificant, like our applause) has to appear immediately after a desirable behaviour of the patient. Our applause immediately after the first trot in saddle is much more important for our patient, than a promise of giving him a favourite toy after the session.

Example 3 - positive reinforcement

Olga, 5 years old, mild mental retardation

Olga has been interested in a group of riders performing some exercises on the meadow. Therefore, we appointed that after she will execute three exercises, we would immediately ride up to look at their training for a moment. Olga could receive desirable reward immediately after executing three exercises, which she doesn't like. In described situation, I have used:

- **Operant conditioning** – because expected patient's behaviour was immediately followed by positive reinforcement (reward), thus next time patient more willingly repeated that behaviour (exercises).

- **Premack principle** – because preferred patient activity (watching raiders on the meadow) has been used as a reinforcing factor for not preferred exercises.

The second strategy based on **operant conditioning** is the strategy of **extinction**. During this therapy the patient does not receive a desirable reward after negative behaviour, (behaviour which we want to become extinct). Unrewarded behaviour is smothered and disappears. Reward can increase frequency of some patient's behaviour, in the same way that a lack of reward expected by a patient can decrease frequency of that behaviour. The easiest way to use this strategy in hippotherapy sessions is just to ignore the negative behaviour of the patient. Ignoring that behaviour, we do not give him expected reinforcement (our attention).

Example 4 - Extinction

Zuzia, 7 years old, sensory integration dysfunction (SID)

Zuzia is a very spoilt child, especially pampered by her father. She is trying to manipulate everyone at hippotherapy sessions as she is used to doing at home. She is pulling faces, pretending that she will cry, lisping and acting like a little child. Since the beginning of our sessions, I have been trying to ignore that behaviour; as a result, after a while Zuzia stopped behaving like that. After just three sessions with this strategy, the number of these negative behaviours (when Zuzia was trying to manipulate me) decreased. Unfortunately, sometimes Zuzia still tries to manipulate me like that and extract something from me...

...that means that this way of behaving is still effective at Zuzia's home, because parents probably reinforce this way of behaving. They are just giving in, yielding to her.

Thus it is extremely important to encourage parents constantly to cooperate, in order not to reinforce negative child's behaviour that we are trying to extinguish during hippotherapy sessions. Put another way, if our activities are opposite to parents' activities... it may cause disinformation of a child instead of improving its behaviour.

Behavioural therapy, due to its simplicity and usage of primary learning's mechanism that are common to people as well as animals, can be used with success in therapy of younger children or children with considerable cognitive deficits. However, as research shows, it is also suitable for healthy or older children as well as for adults.

ARTICLE NINE

The Basis of a Code of Ethics in Riding Therapy - Experiences of Values of Finnish Riding Therapists in Their Clinical Work

AUTHOR: PETRA NOWAK, FINLAND

Article Nine

The Basis of a Code of Ethics in Riding Therapy - Experiences of values of Finnish riding therapists in their clinical work

Abstract of a Scholarly Thesis in a Degree Programme in Riding Therapy (52 pages), Ypäjä Equine College, December 2006

Key words: riding therapy, ethics, professional ethics, code of ethics

The purpose of this thesis is to clarify values of Finnish riding therapists in their practical work. Results may be used in aid to establish the code of ethics of Finnish riding therapy by the Finnish Riding Therapy Association. Thesis is also directed to the Finnish riding therapy students and professionals to give depth to the implementation of clinical work and to awaken to the consideration of ethics in therapy.

Ethics describes and justifies right and good ways to live and function in the world among other people. Ethics are formed by values, ideals and principles that deal with good and evil, right and wrong. The purpose of ethics is to help to make choices. *Professional* ethics ponders the question what kind of action is acceptable when practicing in a profession. It does not differ from other ethics, but it focuses on ethics in actions made professionally. Also questions of professional ethics can lead to different answers.

Several occupations have a code of ethics, which are written commandments of how to operate. Professionals are morally obliged to follow these. They are not bound in law and no-one can order to follow them. A code of ethics is part of work in healthcare and rehabilitation. In riding therapy therapists follow the principles known from their basic profession. In riding therapy he (the therapist) should consider not just himself and the client, but the horse and the helper which have their role in the code as well.

In this thesis the values Finnish riding therapists have were surveyed. The survey had four parts, the claims regarding therapist, client, horse and helper were expressed. The answer was to be given with an argument provided for each choice. On the grounds of these arguments (reasons) common opinions and experiences associated with values in practical work were noted.

This thesis could provide the basis of the code of ethics in Finnish riding therapy.

Article Ten

User Interviews about Therapeutic Riding with Adults with Learning Difficulties

Author: Susanne Hecht, Germany

ARTICLE TEN

USER INTERVIEWS ABOUT THERAPEUTIC RIDING WITH ADULTS WITH LEARNING DIFFICULTIES

Key words: therapeutic riding, user interviews, adults with learning difficulties

Therapeutic Riding can be a motivating movement activity for adults with learning difficulties. From this background I did several interviews, in which I talked to adults with learning difficulties. The main objective of my study was the active inclusion of the target group into the process of research. It was important to me to not only talk about, but also with adults with learning difficulties.

Over the period from February to March 2008 I did 31 interviews with adults with learning difficulties between the ages of 20 and 70 years. All those user interviews were done in the "Centre for Therapeutic Riding" in Lünen (near Dortmund/Germany).

A questionnaire was developed especially for the study, which considered the specifics of the "interrogation" of the target group. Based on the central questions of the study, the questionnaire contained three main parts:

- Reasons for the participation in therapeutic riding and expectations/aims of adults with learning difficulties
- Mood before, during and after therapeutic riding and effects of therapeutic riding
- Satisfaction and wishes of variation at therapeutic riding

If you are interested in the study (detailed planning, realisation and results of the user interviews or perspectives for further research) feel free to e-mail me: **susi.hecht@freenet.de**

BIBLIOGRAPHY:

HECHT, S. (2008): Heilpädagogische Förderung mit dem Pferd bei erwachsenen Menschen mit einer geistigen Behinderung. Eine Nutzerbefragung zur Evaluation des Angebots im Zentrum für Therapeutisches Reiten der Werkstätten der AWO Dortmund GmbH. Unveröffentlichte Bachelorarbeit. Technische Universität Dortmund. Fakultät Rehabilitationswissenschaften. Lehrgebiet Bewegungserziehung und Bewegungstherapie in Rehabilitation und Pädagogik bei Behinderung.

HECHT, S. (2009). „Weil mir das Reiten Spaß macht!". Planung, Durchführung und Ergebnisse einer Nutzerbefragung zum Thema „Heilpädagogische Förderung mit dem Pferd bei erwachsenen Menschen mit einer geistigen Behinderung". In: Mensch und Pferd. Internationale Zeitschrift für Förderung und Therapie mit dem Pferd. Jahrgang 1, Heft 2, S. 70-79. München: Ernst Reinhardt Verlag.

ARTICLE ELEVEN

Leisure Educational Activity Assisted by Horses as Introduction to Therapy

CO-AUTHORS:

LORENZO LUCARELLI, MONTSE CUESTA, MARÍA BARAJAS, SPAIN

ARTICLE ELEVEN

LEISURE EDUCATIONAL ACTIVITY ASSISTED BY HORSES AS INTRODUCTION TO THERAPY

Keywords: Leisure educational activities, Psychological assessment, Leisure as introduction to therapy, Self administration towards self/social integration, Global approach, Health and education

Nowadays in Catalonia many riding clubs offer a hippotherapy or equestrian rehabilitation service; 'though none, offer psychological assessment, and very little educational therapy is provided in the hippotherapy centres.

Therefore we decided to organise educational and leisure group events for psychic disabled as an introduction to the further organisation of therapeutic activities, which we hope to have recognised by our Government in a reasonably short time (few years), since in other European Countries they are already integrated in to the Public Health System.

Our interdisciplinary team was composed of a psychologist, a nurse and a therapeutic riding instructor. We also wanted to integrate into the team a social educator and a physiotherapist, but we couldn't afford so many professionals, so we received the voluntary help of some physiotherapy students, and the role of the social educator was played by the therapeutic riding instructor who is ending his specialisation in ethics, together with the psychologist who has several years experience working in the integration and social education field.

We organised two group experiences of an example therapy session for psychic disabled with 16 people affected by different mental disabilities: Downs syndrome, cerebral palsy, brain damage and others, making the disabled clean and dress the horses, lead them by the rope, and then ride them moving straight, to the sides, upwards and downwards for a few hundred meters.

Following the first two experiences we started a cycle of 6 sessions of psychological treatment assisted by horses for 4 of the 16 participants, who expressed the will to follow our therapy. They do one session per month, so the final outcome is not yet known, since the programme is not yet ended.

What we mean to report here is that in the leisure sessions we noticed a great effect on the behavior of all participants: they all increased their level of activity and communication, they all reacted positively to the experience and tended to be better and more cooperative within the group activities not just during the experience in the riding club, but also the following few days back at the institution where they meet daily.

Furthermore, 25% of the participants have felt and understood the positive effect of the riding session so much that they have asked to do it regularly and in depth.

We have reached the conclusion then that the leisure experience is to be considered a good way to involve psychical disabled people into a horse assisted therapy programme.

Article Twelve

Hands Up For Hippotherapy: An Investigation into the Use of Hippotherapy to Promote Handwriting Legibility for Children who have Disabilities

Author: Jane Liddiard, Therapy Focus, Australia

ARTICLE TWELVE

HANDS UP FOR HIPPOTHERAPY: AN INVESTIGATION INTO THE USE OF HIPPOTHERAPY TO PROMOTE HANDWRITING LEGIBILITY FOR CHILDREN WHO HAVE DISABILITIES

Key Words: Paediatrics, handwriting, hippotherapy, disability, child, occupational therapy, horse

ABSTRACT

Background: Hippotherapy involves providing therapy treatment using the activity of (horse) riding, something that children who have disabling conditions can 'do'. Hippotherapy has been found to assist with developing balance and postural control, attention and motor coordination, which may contribute to the development of skills in the functional area of handwriting, an activity many children who have disabilities have difficulty in 'doing'. This paper investigates the use of hippotherapy to develop the handwriting skills of children who have disabilities.

Method: Seven children, who experienced a variety of disabilities, participated in ten weeks of hippotherapy intervention and a school/home-based handwriting programme. Baseline, midway and post intervention measurements were taken using the Miller Function and Participation Scales (Miller 2006)and the Evaluation Tool of Children's Handwriting (Amundson 1995).

Results: Friedman's Analysis of Variance identified significant change in all areas of the M-Fun, with post-hoc analysis finding improvement after the commencement of hippotherapy in gross motor and fine motor areas. Analysis of the ETCH found significant improvement in word or letter legibility for four of five participants.

Conclusions: Hippotherapy may be a valuable activity to include within a comprehensive occupational therapy program, to develop fine, gross motor and functional skills, including handwriting.

Introduction

We are attracted to stories of people who 'can do'. Individuals who overcome adversity to perform a socially valued role, such as scientist, athlete or businessperson, are held as heroes. Conversely, lower social value is paid toward people who do not achieve valued roles, people who have limited capacities to 'do': this perception impacts on the individual's self worth, and quality of life (Beer, 1998). Occupational therapy seeks to find ways to help people to 'do'. Whether using an engaging occupation to develop a person's capacities of health and function, or through identifying alternate ways for an individual to engage in a valued activity, occupational therapy seeks to improve a person's quality of life (Bracher, 2000). Liz Hartel of Denmark was held in high regard when she won a silver medal in the 1952 Helsinki Olympic Games. Liz overcame the physical limitations imposed by polio to be able to control a horse at a higher standard than most able-bodied persons and has publicly attributed riding horses as contributing to her recovery (Engel, 2000). Closer to home are stories of Australians who have overcome their disabling conditions to challenge able-bodied persons in an able-bodied world. 'Beth' has incomplete paraplegia, is in the Australian paralympic squad and coaches able-bodied persons in dressage http://www.horseriding.org.au/RDAA/category133_1.htm.

Consider 'Ben': born at 28 weeks gestation and slow to develop his skills. Ben has ataxic cerebral palsy, resulting in difficulties walking and running, problems with attention, learning and with participating in school activities. Ben's occupational therapist used the sensory-motor and psychosocial experience of being on a horse to develop Ben's posture and coordination and to improve his ability to focus and concentrate. Ben's improved motor control and concentration was demonstrated in a marked improvement in his handwriting; his ability to follow instructions and independently complete class work; to join in on the playground with his peers; and to talk about 'his' horse, to the envy of his friends.

It is estimated that 12-21% of typically developing children experience difficulties with handwriting, which may involve poor formation, size or alignment of letters impacting on neatness and legibility, and slow speed; a number of children who have diagnosed disabilities, including less physical conditions such as Attention Deficit Disorder and Autism, experience significant issues performing handwriting tasks at school, impacting on participation in class, completing writing tasks and contributing to negative perceptions of competence and self-esteem (Graham & Weintraub, 1996). For a child who has a disability the major tasks in the role of 'student' – to learn reading, writing, numbers and ideas, to participate in sports and play, and interaction with teachers and peers – may be difficult to 'do'. Success at school can have a significant impact on a child's development into adults, their ability to pursue further study or to perform valued roles such as worker and homemaker (Goldstand, Koslowe & Parush, 2005).

Disabling conditions that contribute to handwriting difficulty may involve biomechanical problems, including impaired range of motion, muscle strength, tone and joint stability; sensory processing impairments, such as awareness of body position in space for automatic correction of posture and coordination of upper limb and hand movements; and cognitive impairments, impacting on attention, concept development and motor learning (Pehoski, 2006; Graham & Weintraub, 1996). Intervention to address handwriting cited in research includes demonstration, verbal and cognitive strategies, remediation of biomechanical impairments, motor learning, engagement in sensory-motor tasks and environmental or task modifications (Graham & Weintraub, 1996; Feder, Majnameer & Synnes, 2000; Case-Smith, 2000).

Children who are born with disabilities may have experienced therapy from an early age, and can associate therapy with discomfort, frustration or failure (Macauley & Gutierrez, 2004; Haydon, 1994, Oliver, 1989). Performing therapy away from the clinic or classroom, while participating in an enviable activity, can foster necessary motivation and engagement to attempt challenging tasks and learn skills; particularly if (horse) riding is an activity the child would like to pursue as recreation or sport (Pehoski, 2006; Macauley & Gutierrez, 2004; Haydon, 1994). Hippotherapy is the term used when an Occupational Therapist, Speech Pathologist or Physiotherapist incorporates a horse into a treatment programme in order to improve a client's functioning in everyday tasks (Macauley & Gutierrez, 2004; http://www.ridingdevelopsability.com/hippotherapy, 2007). Hippotherapy provides a whole body experience, engaging all of the senses in a repetitive and graded motor activity that requires an active response, thus optimising neural reorganisation for skilled performance (Pehoski, 2006; Hornacek, 2005).

Hippotherapy involves having a child positioned on a moving horse. Whether the child is directed to sit astride the horse, turn backwards, lie over the horse, kneel up or lie along it's back, the sensory experience is varied: with greater or less vestibular, proprioceptive, tactile and visual input (Heine, 1997). The speed and direction of the horse can be altered so that the sensory experience is more alerting or more calming – promoting attention and focus – and challenges the child to adapt their own movements to maintain balance and symmetry by integrating visual, body position and balance information. The assumption of positions can be utilised to normalise muscle tone in order to improve strength and postural control, and activities such as reaching, grasping and placing objects or singing games can further challenge the child to develop their visual-motor coordination, strength and range of motion (Heine, 1997; Hornacek, 2005).

Hippotherapy is indicated in the development of several functional abilities. Casardy & Nichols-Larsen (2004) identified 10 weeks of hippotherapy as having a positive impact on gross motor function and overall self-care, mobility and socialisation in 10 children with cerebral palsy. Haehl, Guiliani & Lewis (1999) identified an improvement in postural coordination in two children with cerebral palsy, who were seen to approximate an experienced rider's postural control synergy by the completion of the 12-week hippotherapy program and displayed natural patterns of movement reorganisation indicative of trying and refining new movement strategies. Hammer, Nilsagard, Forsberg, Pepa, Skargren & Oberg (2005) identified a relationship between improved balance, self confidence and performance of activities of daily living in adults with multiple sclerosis who participated in 10-11 sessions of hippotherapy. Macauley & Gutierrez (2004) and Macauley (2007) have suggested that the hippotherapy environment and sensory-motor activity stimulates a child's arousal and attention, facilitates interest in interactions and therefore communication with others and promotes a positive affect. These factors were utilised to teach communication behaviours and achieve the speech and language goals of three children with language-learning disorders (Macauley & Gutierrez, 2004) and 15 children with autism (Macauley, 2007).

Studies suggest a relationship between the improvement in functional performance and development in biomechanical, sensory, motor learning, communication and socio-emotional areas through hippotherapy. There are no published studies demonstrating the impact of hippotherapy on handwriting; however, Liddiard (2007) presented a pilot to this research at the American Hippotherapy Association Conference in which four children with disabilities, including 'Ben' described above, participated in hippotherapy as an adjunct to a direct and consultative occupational therapy program. All four participants demonstrated improvements in handwriting legibility, letter size, line regard and pencil control, that were suggested to be associated with improvements identified in the biomechanical and sensory-motor areas of postural control, upper limb control and balance.

Occupational therapists help individuals to perform valued occupations, as well as using occupation as a therapeutic medium. Riding a horse is a valued occupation that can be used to address a range of impairments that frequently impact on children's ability to perform handwriting tasks, influencing their participation in the valued role of 'student'. This study investigates the use of hippotherapy on the functional handwriting of seven students experiencing a range of disabling conditions.

METHOD

Design

A multiple single subject experimental design was selected for the study as the participants vary in age and diagnosed disability and the hippotherapy treatment is individualised according to the specific difficulties contributing to each participants' handwriting problems. This design enables demonstration of change within individuals in a non-homogenous group, while accounting for factors such as maturation that may be seen in longitudinal studies, and is considered a highly appropriate research design for measuring the clinical applications of treatment (Domholdt, 2000; Ottenbacher & York, 1984). An ABA, baseline-treatment-withdrawal design was selected to further enhance internal validity (Domholdt, 2000). Ethics approval was granted through the Edith Cowan University Human Research Ethics Committee.

Participants were assessed using the Miller Function and Participation Scale (M-Fun) (Miller, 2006) five weeks apart for baseline assessments, followed by ten weeks of hippotherapy intervention with a mid-way assessment after 5 weeks, then two post intervention baseline assessments five weeks apart. The participants' teachers were requested to submit handwriting samples on a fortnightly basis for the term of the study. This enabled measurement of change in functional handwriting performance. One participant missed one hippotherapy session due to illness.

Immediately following the first baseline assessment, participants were provided with a handwriting program, focused on letter formation. This consisted of six 10-minute activities per week that could be performed at home or at school and incorporated gross and fine motor activities focused on sensori-motor practice of letter formation, with verbal cuing, based on Peggy Lego's Art of Handwriting (Lego, 1983). The provision of task-specific practice is integral to handwriting skill development and the aim of this programme was to ensure that all participants experienced a similar level of handwriting instruction. Participants performed this programme from the initial baseline assessment until the final baseline assessment, to differentiate between changes in handwriting skill that may be attributable to the handwriting programme from those attributable to the hippotherapy intervention.

Participants

Study participants were clients of Therapy Focus Inc: a not for profit organisation that provides therapy and professional services to children with disabilities in the metropolitan area of Perth, Western Australia for children who have disabilities. This study was advertised through occupational therapists within the organisation and in the organisation newsletter.

Participants were required to demonstrate deficits in postural control, balance and visual-motor control contributing to poor handwriting legibility. Informed consent was sought from parents, and the study was explained to the participants themselves, with informed consent obtained by a primary therapist or non-family member who was familiar with the participant and their favoured communication style and level of comprehension. All participants had a 'Riding for the Disabled' Medical form completed by their general practitioner to ensure that participants did not experience conditions that were contraindicated for (horse) riding,

Children were excluded if there was evidence of visual-perceptual deficits impacting on ability to copy basic shapes and letters, if children were not willing to attempt to copy letters or did not have the cognitive or physical capacity to write letters. In addition, participants were required to refrain from participating in direct occupational therapy intervention for the duration of the study. Eleven children were initially assessed; one child was excluded as he could not copy basic pre-writing patterns or write his name, and a second child withdrew, as her family did not want to discontinue private occupational therapy. Two participants, who completed the intervention, had initial Evaluation Tool of Children's Handwriting (ETCH) (Amundson, 1995) handwriting legibility scores (both word and letter legibility) greater than 80% and were thus excluded from the data analysis.

Participants ranged in age from seven- to twelve-years-old, included five boys and two girls, and the disabling conditions included intellectual disability, dyspraxia, autism spectrum disorders (2 participants), cerebral palsy, Down's syndrome and Mobius Plus syndrome. Participants did not participate in direct occupational therapy for the duration of the project, but had previously received a consultative service through Therapy Focus that may have recommended use of particular equipment or strategies, and it was requested that these strategies remain constant throughout the study. Two of the participants participated in a weekly gross motor program and one participant attended weekly 'Riding for the Disabled' (horse) riding sessions that continued for the duration of the project.

Intervention

Hippotherapy intervention took place at one of two locations within the metropolitan area: one an existing 'Riding for the Disabled' (RDA) centre; the other at an education support school, with RDA coaches and volunteers transporting horses and equipment to the venue for the day. Participants attended ten (weekly), 45-minute hippotherapy sessions, with one participant missing one session due to illness.

Horses, horse handlers and sidewalkers were trained and accredited through RDA-Australia Hippotherapy, including emergency procedures. Riders were matched to their horse based on the degree and quality of movement, and the availability of the horse on each day. Where possible, the horse and volunteers were kept consistent for each rider.

Hippotherapy intervention varied according to each participant's particular difficulties and graduated according to their confidence, strength and development of balance. In addition to the usual forward-astride position, participants were often required to move into backward sitting position, optimising pelvic alignment and decreasing visual information for balance. In this position, fine motor and writing activities were implemented, such as rotating an instruction cube (to tell the horse to go faster, slower or stop) or writing with chalk on the horse's rump. To develop trunk strength, participants were required to lie prone over the horse's barrel or assume a quadruped position along the horse's back, with activities such as reaching forward or sideways used to challenge strength, coordination and balance. Participants were also supported to sit sideways on the horse, to kneel or stand on the horse (in later sessions), challenging balance and trunk control, and developing abdominal and back strength and confidence. Coordination activities in this position included reaching to apply hair clips to the horse's mane or throwing magnetic darts at a dart board. The horse's movement was also used to help the rider to focus or to challenge their balance and strength. Rapidly changing speed or direction challenged attention and promoted active engagement in the session activities. Changes in pace were also utilised to develop abdominal and trunk strength, and variations in direction to develop symmetry, lateral strength and balance.

Measures

The M-Fun (Miller, 2006) was utilised to assess visual motor control, fine motor abilities and gross motor abilities for baseline, mid-intervention and post intervention measures. This is a standardised occupational therapy assessment for children aged 4 to 8 years. This assessment incorporates functional skills, including handwriting, appropriate to the developmental and skill range of the participants and reduces testing issues due to language or cognitive deficits through supported task practice. Pre-test 2 and mid-test were conducted by an occupational therapy student (trained in using the M-Fun and who participated in conducting pre-test 1) for two of the participants.

The M-Fun data progress scores were used for comparison of scores over time. These are mapped using an item response theory model and have a mean of 500 with a standard deviation of 100 (Hambleton, Swaminathan and Rogers, 1991, cited in Miller, 2006, p99). Each area of the assessment, Visual Motor, Fine Motor and Gross Motor, was analysed using the SPSS 15.0 package. Friedman's analysis of variance was used to identify differences in scores in the five test conditions, with an alpha level at $p>0.05$. Post Hoc analysis was conducted using Wilcoxon Signed Ranks Test, to identify differences between each test condition, with an alpha level set at 0.15, with Bonferroni adjustment for repeated measures (Domholdt, 2000).

The Evaluation Tool for Children's Handwriting (ETCH) (Amundson, 1995) is a commercially available criterion referenced handwriting tool, appropriate for comparing legibility scores of several writing samples by one subject. The ETCH scoring considers important components of text legibility: formation, shaping, proportion, spacing and overall legibility of a letter or word out of context; and has moderate reliability, though studies have questioned its validity in relation to teacher's handwriting legibility ratings (Tomcheck & Schneck, 2006; Susawad, Trombly, Henderson & Tickle-Degnen, 2002). Classroom handwriting samples were assessed according to the criteria of the ETCH. Samples were coded both for date and participant to enable blind analysis, and each scored separately by three scorers to promote reliability of results. The scorers underwent the training protocol, specified in the ETCH manual. Consideration was taken whether samples were identified as near-point copying, far-point copying, writing from dictation or composed, and scored according to the criteria for each circumstance.

One participant's handwriting samples were traced words, and could not be assessed utilising the ETCH, and two participants did not have sufficient samples provided to establish a baseline. The type of samples provided by the teachers were not all consistent (creative writing, copying from the blackboard, worksheets or a brainstorming activity). Scores from the ETCH were averaged and graphed using Microsoft Office Excel 2007. Visual analysis and two-band standard deviation method was utilised to determine significance of change from the baseline to the intervention phase (Dumholdt, 2000; Ottenbacher & York, 1984; Ottenbacher, 1986).

RESULTS

The mean scores for each test condition and each area of the M-Fun are depicted in Figure 1.

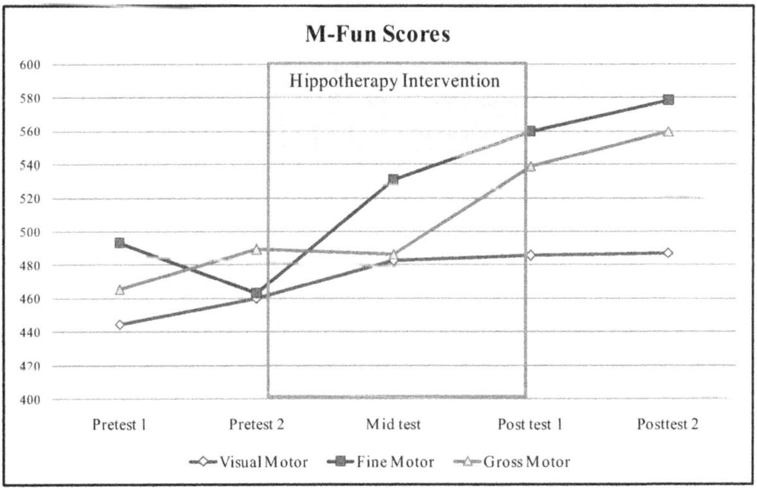

Figure 1. Average M-Fun progress scores

Friedman's analysis of variance identified a significant difference in Visual Motor (x^2 = 11.066, p<0.05), Fine Motor (x^2 = 22.302, p<0.000) and Gross Motor (x^2 = 16.800, p<0.05) scores, between test conditions. Post hoc analysis identified a significant difference (p<0.037 after Bonferroni adjustment) for Fine Motor scores between pre-test 2 and mid-test (z = -2.366), and for Gross Motor scores between mid-test and post-test 1 (z = -2.117). No significant difference was identified between each test condition for Visual Motor scores, despite the statistically significant result using the Friedman's analysis of variance; however, a large difference was identified between pre-test 2 and mid-test. Of the handwriting samples analysed, three participants demonstrated a significant change in either word or letter legibility and one participant did not show a significant change.

DISCUSSION

The results from this study indicate that the inclusion of hippotherapy into an occupational therapy intervention programme may be effective for the development of fine and gross motor skills and functional handwriting performance. That this may be due to improvements in areas of postural control, upper limb coordination, visual motor coordination and attention to task; however, the functional tasks evaluated in the assessments utilised can only reflect possible changes in these underlying areas.

Significant change was identified in both fine and gross motor abilities on the M-Fun, after the introduction of hippotherapy. Change in gross motor abilities in the second half of the intervention phase, but not the first half, is consistent with a child's neural re-organisation of motor patterns after several weeks of intervention. This was described in Haehl et al's (1999) study of two children with cerebral palsy, who demonstrated an initial decrease in postural control, followed by significant increase, and may reflect patterns identified in typically developing children as they develop from being dependent on visual information for postural strategies to being able to use somato-sensory information (Westcott & Burtner, 2004). The M-Fun gross motor tasks, including single leg stand, ball skills and jumping may have been influenced by postural control, but other aspects of coordination such as visual motor abilities may also have contributed.

On average, participants' fine-motor scores decreased between the first and the second pre-tests, leading to a greater increase by mid-test. Changes in fine motor abilities would be expected to follow gross motor, as motor control theories demonstrate gross motor ability generally precedes refinement of skills. As the hippotherapy sessions incorporated fine motor activities (such as using pegs, rotating blocks, using small pieces of chalk) it may be that the activities incorporated into the therapy sessions contributed to the development of fine motor skills.

The Friedman's analysis of variance identified a statistically significant difference in visual-motor scores and, although the Wilcoxon Signed Ranks test did not find a significant difference between any test conditions, an improvement can be identified between pre-tests 1 and 2 and a larger improvement between pre-test 2 and mid-test. Introduction of the handwriting program after pre-test 1 may have contributed to the change of scores between pre-test 1 and 2. The intervention (handwriting program and hippotherapy) had a significant change on the visual-motor skills of the participants; improvement accelerated after the introduction of hippotherapy, with small improvements continuing in the second half of the hippotherapy intervention and skills maintained after the hippotherapy was completed.

The ETCH was chosen to identify the functional outcome of improved motor control in typical classroom writing tasks, with a meaningful change found for three out of four participants. Aspects of handwriting that impact on ETCH scores include orientation of letters to the line, spacing between words and letters, and letter formation: these may be influenced by memory and learning. Letters or words were given a negative score if any of these aspects were not met, which decreases this assessment's sensitivity to improvements in letter formation (due to improved pencil control) and can create inconsistencies between letter and word legibility ratings; for example, letters of a word may be formed correctly and given a positive score; but if spacing was insufficient between words, the words would be given a negative score. This insensitivity to changes in handwriting ability may account for reports that ETCH scores show a low correlation with teachers' perceptions of handwriting ability (Tomcheck & Schneck, 2006; Susawad et al, 2002).

In this study, there appeared to be an improvement in some aspects of legibility in every-day handwriting. Studies into the effectiveness of hippotherapy as a treatment activity have frequently identified a meaningful improvement in functional skills, without statistically significant change in body impairments (Casardy & Nichols-Larsen, 2004; Hammar et al, 2005). Increased confidence in movement activities (fine and gross motor) may generalise to a willingness to attempt movement activities in other areas, such as the concurrent improved balance with increased self-confidence and improvement in performance of daily living tasks, reported in adults with multiple sclerosis (Hammar et al, 2005).

Increased self-confidence and esteem, fitness and social benefits have been associated with engagement in a physical activity, particularly for children who have disabilities, whose physical activity levels are typically lower than non-disabled peers (Downs & Briffa, 2004). Hippotherapy is a community-based and age-appropriate recreation activity that may also have these psycho-social benefits of increased confidence and motivation in school performance.

Handwriting is a complex activity, with difficulties requiring intervention ranging from remediation of underlying neurological impairments such as strength, visual perception or tactile awareness; through development of skills such as visual motor control; to adaptation of the task or environment to optimise success. This study focused on remediating selected neurological impairments and developing skills; however, did not address the child's environment or the handwriting task. The previous study into hippotherapy and handwriting (Liddiard, 2007) identified improvements in aspects of handwriting related to motor control: pencil control, pencil pressure, letter formation and letter size, with the functional outcome of improved readability.

This current study supports the suggestion that hippotherapy may be effectively utilised to develop skills that contribute to handwriting performance; however, occupational therapy literature promotes engagement in functional activity as an integral part of therapy for that occupation. Intervention for skills and underlying impairments may contribute to performance but function is a result of interactions between the person, the task and the environment. Thus, intervention that does not consider all aspects will be limited in its outcome, and it is recommended that if hippotherapy be utilised it is to compliment a comprehensive occupational therapy program that incorporates all if these aspects.

LIMITATIONS

Results in this study must be considered with caution, as there were large changes in baseline scores for many participants in both the M-Fun and the ETCH assessment. This may be reflective of the complex nature of functional skills such as handwriting, where aspects such as the nature of the particular task, time of day, time of the year, motivation (behaviour management), or environment may vary greatly between participants; for example, one participant was noted to have issues with medical control of epilepsy during the study, which was reported to impact on motor control and attention to task.

Difficulties have been documented in the use of standardised assessment with children who have disabilities and behaviour management issues were identified for some of the students, with incentives (food or activity) and encouragement required to varying degrees to encourage some participants to complete the assessment tasks: in these cases, performance is likely to be less than optimal.

The chief investigator and two occupational therapy students were involved in both the assessments and the intervention. Steps were taken to promote validity of the ETCH scores through using three scorers and blinding to date and participant. Continued improvement in M-Fun scores may be due to participants becoming familiar with the investigator during the hippotherapy sessions, enhancing willingness to comply with tasks. In addition, three assessors conducted the assessments, each of whom had participated in scoring the initial assessments and practiced administering the assessment. Consistency in testing conditions is integral to the reliability of a single case design.

Future studies may benefit from increasing the baseline phase to establish a more consistent pre-intervention handwriting sample and to decrease the impact of outside influence. Increased occupational therapist involvement with schools may have enhanced the implementation of the skill activities in the handwriting program and increased the submission of handwriting samples. Further suggestions include using a paired control group, investigating a more homogenous population and to include measures of participation and quality of life.

Conclusion

This study provides further support for therapists to consider utilising hippotherapy in treatment programs for children who experience disabling conditions. Hippotherapy may be used as an intervention activity with suggested benefits to clients from the sensory-motor experience, the development of strength and normalisation of muscle tone, attention and engagement, confidence and motivation to attempt motor tasks. In addition, traditional therapy activities can be incorporated into the session, such as writing with chalk, completing puzzles or even including dressing skills or appropriate social interactions. The inherent qualities of hippotherapy can assist a child in overcoming physical or psychosocial barriers to occupation and promote further development of occupation-specific skills: hippotherapy may allow a child to 'do' everyday tasks, while the child feels motivated and enjoys the activity they are doing. This is particularly pertinent to a child or family who is coping with the impact of disability on everyday living, who may be weary of a lifetime of therapy and who gains increased inspiration to participate and achieve life goals in this more normalised experience (Macauley & Gutierrez, 2004; Heine, 1997).

The most resounding outcome is that, although measured changes at the level of body function and structure are subtle, the impact of hippotherapy at the activity and participation levels (ability to perform valued tasks such as handwriting) is more valuable.

In Australia, Hippotherapy is conducted under the umbrella of the 'Riding for the Disabled Association'. Therapists who wish to utilise hippotherapy with their clients should undertake training with 'Riding for the Disabled' and link in with their local centre.

FIGURES AND TABLES:

Figure 2. ETCH scores for four participants showing baseline mean and two standard deviation band (indicated over intervention phase) for word ratings and letter ratings.

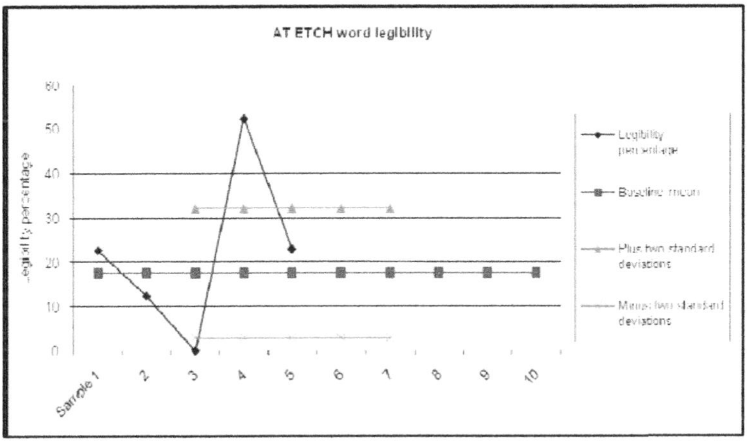

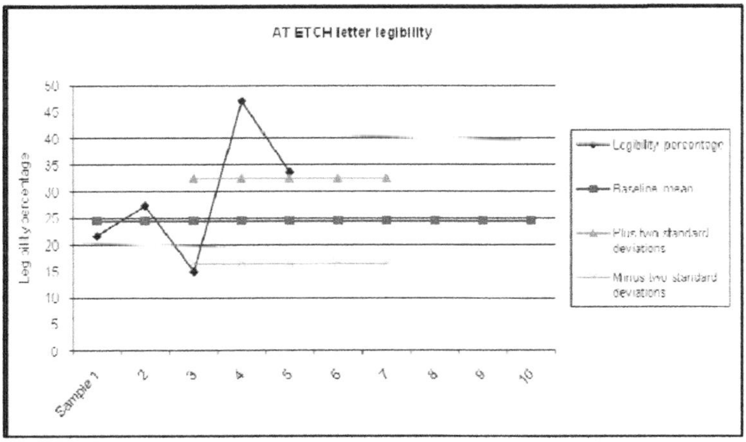

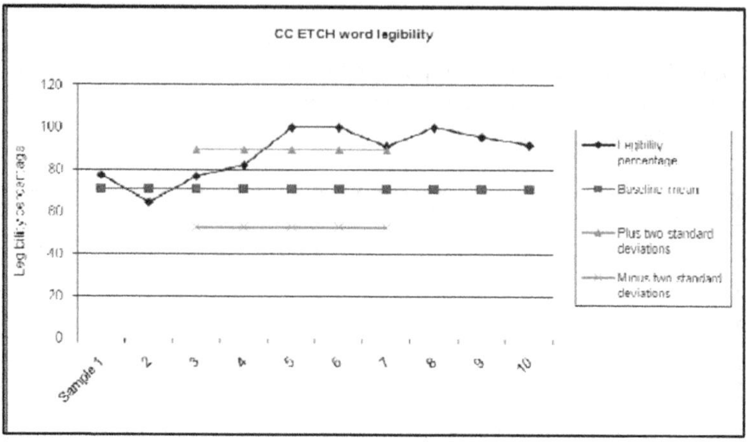
CC ETCH word legibility
120
100
80
60
40
20
0
Sample 1
2
3
4
5
6
7
8
9
10
Plus two standard deviations
deviations

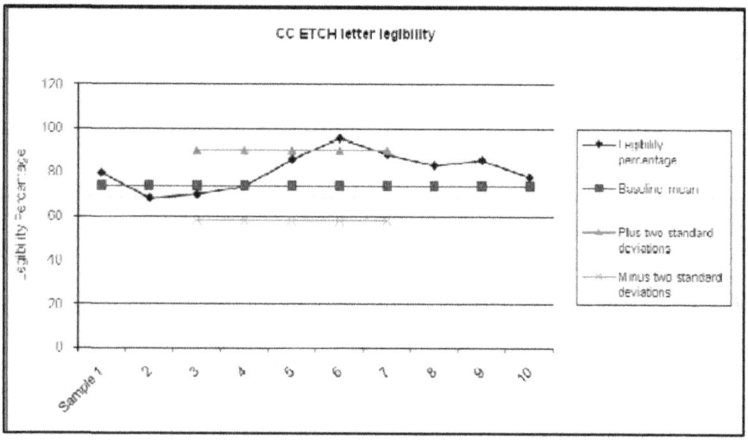
CC ETCH letter legibility
120
100
80
60
40
20
0
Sample 1
2
3
4
5
6
7
8
9
10
Plus two standard deviations
Minus two standard deviations

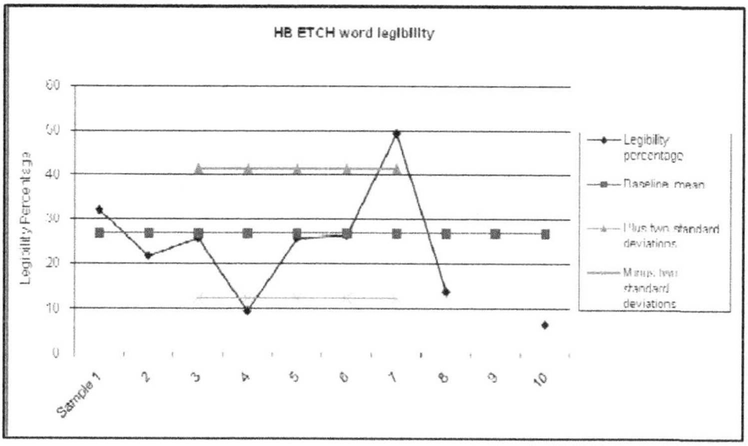
HB ETCH word legibility
Legibility Percentage
Legibility percentage
Baseline mean
Plus two standard deviations
Minus two standard deviations
Sample 1

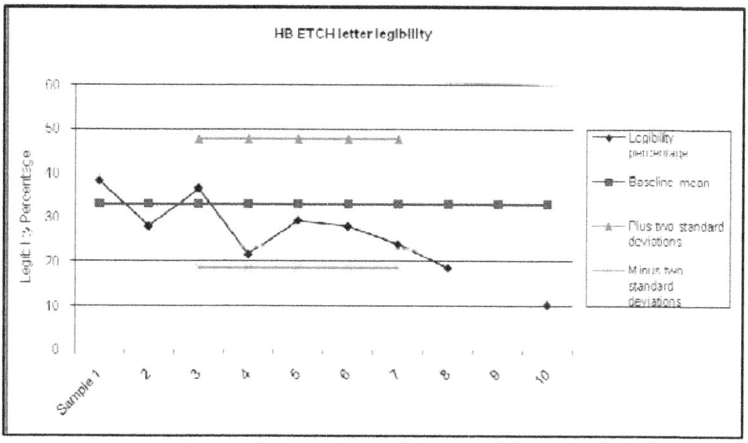
HB ETCH letter legibility
Legibility Percentage
Legibility percentage
Baseline mean
Plus two standard deviations
Minus two standard deviations
Sample 1

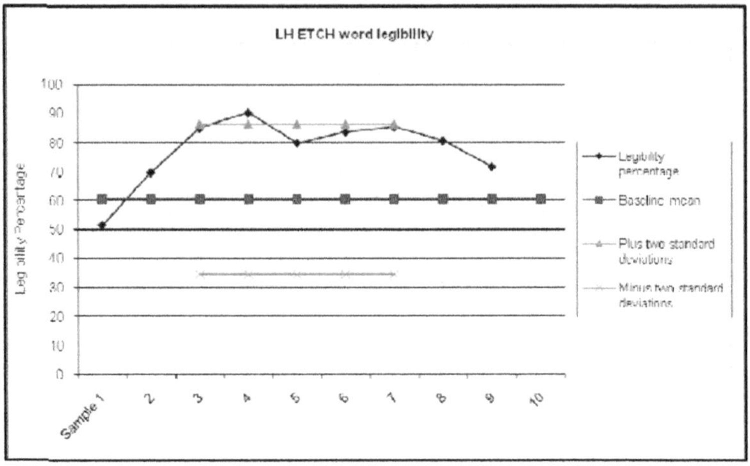

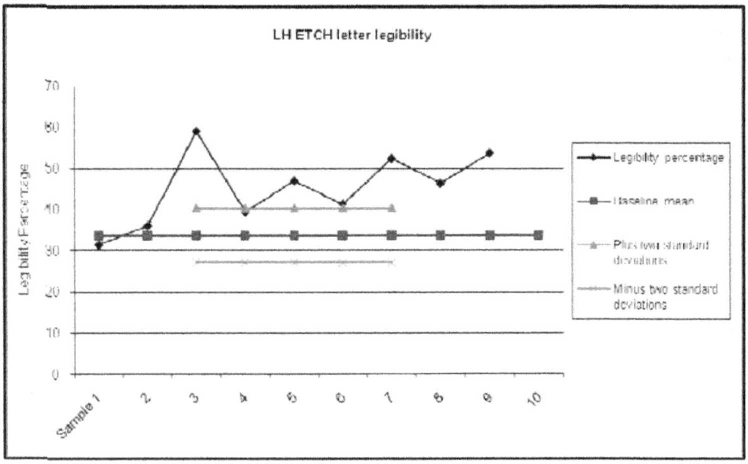

ACKNOWLEDGEMENTS:

This research project was made possible by a research grant from the Non Government Centre for Support. Many thanks to all those from Riding for the Disabled WA, Therapy Focus and Edith Cowan University for their support, particularly Janet Richmond.

REFERENCES:

AMUNDSON, S.J., (1995). *The Evaluation Tool of Children's Handwriting*. Homer Alaska: OT Kids

BEER, D.W., (1998). The illness and disability experience from an individual perspective. *Willard & Spackman's Occupational Therapy (9th Ed)*. Philadelphia: Lippincott-Raven Publishers.

BRACHER, M., (2000). Therapeutic horse riding: what has this to do with occupational therapists? *British Journal of Occupational Therapy*. 63(6): 277-282

CASARDY, R.L., & NICHOLS-LARSEN, D.S., (2004). The effect of hippotherapy on ten children with cerebral palsy. *Pediatric Physical Therapy*. 16: 165-172

CASE-SMITH, J., (2000). Effects of occupational therapy services on fine motor and functional performance in preschool children. *American Journal of Occupational Therapy*. 54:372-380

DOMHOLDT, E., (2000). *Physical Therapy Research*. USA: WB Saunders Co.

DOWNS, J., & BRIFFA, T. (2004). Levels of physical activity in children with disabilities: a review of the literature and mapping study of organisations in Western Australia. Perth: Therapy Focus Inc.

ENGEL, B.T, (2007). Hippotherapy in the practice of occupational therapy. In: B.T. Engel & J.R. McKinnon (Eds). *Enhancing Human Occupation through Hippotherapy: A Guide for Occupational Therapy.* Bethesda, M.D.: AOTA press

ERHARDT, R., & MEADE, V., (2005). Improving handwriting without teaching handwriting: the consultative clinical reasoning process. *Australian Occupational Therapy Journal*. 52(3):199-210

FEDER, K., MANJNEMER, A., & SYNNES, A., (2000). Handwriting: current trends in occupational therapy practice. *Canadian Journal of Occupational Therapy*. 67(3): 197

Goldstand, S., Koslowe, K. & Parush, S., (2005). Vision, visual information processing and academic performance among seventh grade schoolchildren: a more significant relationship than we thought? *American Journal of Occupational Therapy.* 59: 377-389

Graham, S., & Wientraub, N., (1996). Review of handwriting research: progress and prospects from 1980 to 1994. *Educational Psychology Review.* 8(1): 7-87

Haehl, V., Giuliani, C. & Lewis, C., (1999). Influence of hippotherapy on the kinematics and functional performance of two children with cerebral palsy. *Pediatric Physical Therapy.* 11: 89-101

Hammer, A., Nilsagard, Y., Forsberg, A., Pepa, H., Skargren, E & Oberg, G., (2005). Evaluation of therapeutic riding (Sweden) / hippotherapy (USA). A single-subject experimental design study replicated in eleven patients with multiple sclerosis. *Physiotherapy Theory and Practice.* 21(1): 51-77

Haydon, S., (1994). Horses as a vital link in the development of a person with a disability in an able-bodied world. *Eighth International Therapeutic Riding Congress Complete Papers,* 195 – 197.

Heine, B., (1997). Hippotherapy. A multisystem approach to the treatment of neuromuscular disorders. Australian Physiotherapy. 43(2) 145-149

Hornacek, K., (2005) Basal, though overlooked factors affecting posture in therapeutic riding. *Scientific and Educational Journal of Therapeutic Riding.* p 1-11

Lego, P., (1983) *Peggy Lego's Art of Handwriting.* Australia: Real Books by agreement with Universal Publishing Company

Liddiard, J., (2007). *Hippotherapy All Write!* 2007 National Conference American Hippotherapy Association Inc. Atlanta, Georgia.

Macauley, B.L. & Gutierrez, K.M. (2004). The effectiveness of hippotherapy for children with language-learning disorders. *Communication Disorders Quarterly.* 25(4): 205

MACAULEY, B.L., (2007). Effects of equine movement on attention and communication in children with autism. 2007 National Conference American Hippotherapy Association Inc. Atlanta, Georgia.

MILLER, L.J., (2006) *Miller Function and Assessment Scales.* Texas: Harcourt Publishers Inc

OLIVER, C.E., (1989). A sensorimotor program for improving writing readiness skills in elementary age children. *American Journal of Occupational Therapy.* 44(2): 111-116

OTTENBACHER, K., & YORK, J., (1984). Strategies for evaluating clinical change: implications for practice and research. *American Journal of Occupational Therapy.* 38(10): 647-659

OTTENBACHER, K., (1986). Reliability and accuracy of visually analysing graphed data from single-subject designs. *American Journal of Occupational Therapy.* 40(7):464-469

PEHOSKI, C., (2006). Cortical control of hand-object interaction. In: A. Henderson & C. Pehoski (Eds). *Hand Function in the Child. Foundations for Remediation.* Missouri: Mosby Inc.

SUSAWAD, P., TROMBLY, C., HENDERSON, A., & TICKLE-DENEGEN, L., (2002). Testing the effect of kinesthetic training on handwriting performance in first grade students. *American Journal of Occupational Therapy.* 56: 26-33

TOMCHEK, S.D., & SCHNECK, C.M., (2006). Evaluation of handwriting. In: A. Henderson & C. Pehoski (Eds). *Hand Function in the Child. Foundations for Remediation.* Missouri: Mosby Inc

WESTCOTT, S.L., & BURTNER, P., (2004). Postural control in children: Implications for paediatric practice. *Physical and Occupational Therapy in Paediatrics.* 24(1/2): 5-55

ARTICLE THIRTEEN

Equine Assisted Therapy Camps for Children with ADHD or ADD

CO-AUTHORS:

CARINE DEBANDT & STEFI LEJEUNE,
ARTEVELDEHOGESCHOOL GENT, BELGIUM

ARTICLE THIRTEEN

EQUINE ASSISTED THERAPY CAMPS FOR CHILDREN WITH ADHD OR ADD

As a final project and part of the postgraduate training "Equine Assisted Therapy" (EAT) at the Arteveldehogeschool in Gent, we organised a graduation project of 3 EAT camps for children with Attention Deficit Hyperactivity Disorder (ADHD) or with Attention Deficit Disorder (ADD) without diagnosed mental restriction. The purpose of these camps was the stimulation of the general wellbeing, both mental and physical, of the children. The camps offered a motivating opportunity where they could discover and develop themselves and learn to function in a group.

The camps took place in Riding Centre Dennenhof at Landegem (Belgium), where over a period of 5 days the children were given a combination of psycho-education, behavioural therapy, psychomotor therapy, creative therapy and relaxation exercise. During these days and therapies the children learned to know themselves in a context of different activities with the horses.

During the psycho-education the children learnt more about ADHD in a playful way. During the behavioural therapy they learnt to control their impulsivity during several activities. For this we used the cognitive behavioural therapy method "Joep Loep". This method focuses on resolving problems in the 'now' situation, without raking up the past.

The children also worked creatively with the theme 'horse'. Ample time was also devoted to relaxation on and with the horse.

During the camps we worked with themes. These themes were a red line during the whole camp. We worked for example with themes such as 'Cowboys and Indians', 'Prince and Princess' and 'Circus'.

During the camps the children were guided by a multidisciplinary team. The team consisted of 2 physiotherapists, 1 occupational therapist and 1 psychologist, supported by several volunteers.

Every discipline had its own concrete objectives. These were all brought together through the camp programme which integrated all the objectives.

Children were evaluated daily on the basis of observation data and objectives were adjusted as necessary.

During the development of the camp programme we a positive and successful experience was always sought. In a playful, creative and communicative way the children were confronted with their 'being different'. Negative feelings were discussed, positive feelings were encouraged.

Parents were also involved in the event. They provided additional information about their child by answering a questionnaire in advance. During the camp we worked using a daily booklet that went back and forth between ourselves and the parents in which we all wrote our observations.

Positive results were achieved during these Equine Assisted Therapy camps. The children learned about their own strengths and weaknesses. Through the various activities they learned a lot about ADD –ADHD. They learned to accept and comply with rules, they took care of the horses, they learned to interact properly with each other and, last but not least, they learned to relax.

Notes

Made in the USA
Las Vegas, NV
03 May 2026

46330530R00066